POLICIES AND
POLITICS IN
WESTERN
EUROPE

Policies and Politics in Western Europe

The Impact of the Recession

Edited by F.F. RIDLEY

CROOM HELM
London ● Sydney ● Dover, New Hampshire

320.94
P.766

©1984 F.F. Ridley
Croom Helm Ltd, Provident House, Burrell Row,
Beckenham, Kent BR3 1AT

Croom Helm Australia Pty Ltd, First Floor,
139 King Street, Sydney, NSW 2001, Australia

British Library Cataloguing in Publication Data

Policies and politics in Western Europe.
   1. Europe—Politics and government—1945-
   I. Ridley, F.F.
   320.94   JN94.A2

   ISBN 0-7099-2280-9

Croom Helm, 51 Washington Street, Dover,
New Hampshire 03820, USA

Library of Congress Catalog Card Number: 84-45284
Cataloging in Publication Data Applied For.

85-7834

Printed and bound in Great Britain
by Billing & Sons Limited, Worcester.

CONTENTS

POLICIES AND POLITICS IN WESTERN EUROPE

Chapter One

INTRODUCTION

F.F. Ridley

The chapters of this book introduce students to
the government and politics of five West European
countries - Britain, France, Germany, Italy and
Sweden - in no order of significance except alpha-
betical. They give a broad account of the sort of
problems these countries face and the policies adopted to meet them. They offer a background on the
substance of politics in the countries concerned and
are intended, at the same time, to serve as windows
on the political systems in question. Though the
volume stands perfectly well by itself, it is also
meant as a complementary textbook, adding an important perspective to those which give fuller accounts
of political forces and government institutions. A
concluding chapter provides an overview of European
experience.
      Each chapter can be read by itself, by students,
for example, whose courses include only some of the
countries covered here, though these have been
chosen to match the countries most frequently covered in comparative courses on European politics.
Though chapters are broadly similar in framework to
facilitate comparisons by readers, the book does not
discuss methodology, elaborate models or, except in
the concluding chapter, lay much stress on cross-
national generalisations. The reason for this is
simply to allow the widest possible use as a source
of information for courses structured in a variety
of ways.
      There are many ways of studying government and
politics. The institutional approach focuses on the
machinery of government: the rules of the system and
its organisation charts. The behavioural approach
focuses on the performance of actors in the system
rather than on the formal arrangements. Marxists
examine the economic bases of power and relate the

political system to socio-economic forces. Historians' accounts, which link political phenomena in chronological order, look very different from the accounts of systems theory which order material in terms of their functional relationships. The policy approach, a relatively new fashion, though its pedigree can be traced back, is yet another way of looking at government - or rather, several different ways since, like most fashionable terms, it means different things to different practitioners.

The attention of ordinary people - those who are not politicians or political scientists - is sometimes caught by the passing drama of politics. It is more likely, however, that they will concern themselves with the substance of policies than with the policy-making process: with what the charmed circle of decision-makers do to them, rather than with what its inhabitants do to each other. Most people, most of the time, probably care less about the *how* and *why* of government action that about the *what*. Some, though not as many as political activists like to think, occasionally discuss the political system: electoral behaviour, the role of pressure groups, the influence of bureaucrats, the socio-economic bases of power. More often, they argue about policy issues: what services the welfare state should provide, how schools should be run, what should be done about youth unemployment or inner-city decay.

Students of policy, however, are not ordinary people either, and their interest is not a direct reaction to the effect policies have on their own lives. Policy studies have come to be recognised as an academic subject: the word 'policy' appears increasingly in the description of courses and the title of books. The label is ambiguous, nevertheless, and the student registering for a course or purchasing a book can never be quite sure what he/she will get. The field of policy studies is populated by writers with quite different concerns. Most political scientists study the policy-making process to explain how it operates, why it operates as it does and why, as a result, its outcomes are as they are. Others, sometimes public administrationists, rarely political scientists, study techniques that can be used *in* decision-making.

Thus defined, policy studies become a vocational subject. Those trained as policy analysts (a trade description more common in America than Britain so far) contribute to the formulation of policy as advisers to government or, as officials, within government. Their expertise includes a range of tech-

niques (some would say a rag-bag since they have no
common disciplinary base), of which cost-benefit
analysis is probably the widest known, designed to
improve policy-making as a rational process. Put
another way, this is not the explanatory approach of
political science but a prescriptive approach: rec-
ipes for good management. Its advocates, indeed, have
been called missionaries from management science.
Unkind perhaps, but just as missionaries overlook
the wisdom of other faiths, so management scientists
may overlook the contribution of other disciplines.
A study of policy-making techniques which ignores
political science makes no practical sense. Policy
analysts need to understand the political system
which interacts with all stages of their work: set-
ting the goals, imposing constraints on the means to
achieve them, and distorting rational choice even
within that framework.

For many political scientists, on the other
hand, the policy approach is just another way of
looking at the political system. The political-
system cake is much the same as in other approaches,
but it is sliced in a different manner. Instead of
organising material around parts of the system
(chapters on voters, parties, pressure groups, parliament, cabinet, civil service), it is organised
around fields of government action (economic policies, welfare programmes, educational reform, etc.).
Focus is on the 'patterned flow of issues'. There
are clear advantages to the policy perspective, compared to institutional, behavioural and systemic
ways of looking at government, all of which, in their
different ways, provide the student with a lot of
information about how the political system operates
but often leave him/her with little idea of what it
is that governments actually do. The disadvantage,
of course, is that taken alone it may leave him/her
with no map of the system as a whole. Policy-
orientated books should generally be read in con-
junction with other textbooks on the government and
politics of the countries concerned.

Studying the political process through policies
involves asking questions such as the following. How
has an issue arisen? Has it been put on the agenda
by a party programme or an interest group campaign-
ing for a particular reform, or has it forced itself
on the attention of government directly (e.g. an
international crisis)? How is the problem perceived
by actors within the political system? What are the
different solutions proposed and how do they relate

to group interests or party ideologies? How do these actors make their inputs into the policy-making process? What power do they have and why? How are their claims met in the policy outcome? What other considerations affect the decision-makers, e.g. their own perception of domestic and international problems? How do the institutional arrangements of government, the way the machine itself is organised, affect the transformation of inputs into outputs? What influence do actors within the machine have, notably civil servants? What is the output: what sort of policy emerges? How is policy affected by implementation? To what extent is policy actually shaped by administrative action, detailed rules and the behaviour of bureaucrats? What impact does a policy have? Is it effective in solving the problem as intended or does its implementation create new problems, thus new demands on the policy-making system? And when these questions are answered with regard to a number of policy issues, can one generalise about the system as a whole? Can one describe common patterns? Can one explain why certain outcomes are more likely than others? What theory can be built on policy-field case studies or on country studies of policy-making arrangements?

This is not an exhaustive shopping-list, much less a framework for case studies. What may be noted, however, is that by and large the questions are those of political science. By focussing on them, students will learn a great deal about political institutions, political behaviour and political power. Their understanding of how issues are processed, however, may not be matched by an understanding of the issues themselves or the solutions adopted. This tendency to concentrate on the form rather than the substance is natural enough, since the substance of policies forms part of other disciplines: economics, social administration, education theory, criminology, planning, etc. The policy approach nevertheless injects some understanding of these matters. In other words, it should not simply be another way of analysing the political system; it is also a way of broadening students' horizons, giving them a point of entry into the agenda items of public debate, too often excluded from political science degrees as the province of other academic departments.

Political scientists have always maintained that theirs is a cross-roads field of study, involving constitutional law, history, sociology, philosophy and the like, as well as methodologies they consider

4

their own. Policy studies must be far more inter-
disciplinary than that, however. They can not be
contained in political science if they are really to
make sense. To understand the making of economic
policy, it is not enough to know which political
forces back monetarism, which are Keynesian and
which favour a comprehensive system of economic
planning. Nor will it do simply to note what the
different schools stand for. The same applies to
policy proposals in the fields of health, welfare,
education, housing, town planning, transport, nuclear energy, environmental protection and defence.
One needs to understand something of their *internal*
logic if one is to understand the policy-making
process fully.

   Obviously, students of political science can not
master all these subjects. This poses a risk for
policy studies. Political scientists tend to see the
policy-making process as if it were essentially
political: policy the outcome of a play of power. In
many cases such a perspective is not far wrong.
Britain is governed by monetarist policies because
Mrs Thatcher won the Conservative leadership election and the Conservatives won the general election.
In other fields, or at other times and in other
countries, governments may not be as ideologically
committed, or as strong-willed in the face of
pressures, or they may depend on coalitions: conflict, bargaining and compromise will shape policies.
This, too, falls within the purview of political
science. There are fields, however, where subject
rationality plays a role, where 'technical' argument, evidence and logic, influence decision-makers.
It is true that experts often fail to agree. In the
field of crime prevention, for example, there is
deep-seated disagreement on the causes of crime and
the effect alternative measures will have on crim-
inals. Much of the debate among policy-makers (if
not at Conservative Party conferences) is nevertheless conducted in the language of criminology. If
political scientists focus too much on power, they
may underrate the degree of rational debate that
goes on within the policy-making machine and its
influence on policy-formulation. Policy studies
grounded in political science often tell one little
about the analytic capacity of the machine, the
subject expertise available to it, and how such
inputs are digested. Policy decisions may be less
rational than they should be, but the preparatory
work, civil service briefings for example, involve
non-political considerations that students of

5

politics will find difficult to handle unless they read in other disciplines as well as their own.

There is another important aspect of policy studies which they may overlook. It is not just that one must understand some of the internal logic of an issue, the relevant specialism, to understand the policy-making process fully, but that understanding the process is only part of what policy studies are about. What most people want to discuss is not how decisions are taken but what sort of decisions *should* be taken. Such an approach to policy studies would start with the identification of a problem, list the goals one might want to achieve, and then analyse the various solutions to see which is most likely to produce the desired ends. That, after all, is how civil servants brief their ministers. Although the label begs a lot of questions, one might call it a normative-rational approach. Of course, political factors need to be taken into account: there is little point in advocating a policy that is not politically manageable. This is especially true if it is immediate adoption one has in mind; less important, though still relevant, for those with a longer-term aim of influencing the climate of opinion.

The point here is a simple one. Political scientists may order their studies in a way that is inappropriate for normative inquiry. They start with politics and the policy-making process, and they explain the outcomes. They are often less interested in the range of policy options on offer at the start of the process, including those that failed to get adopted or, at least, to obtain sufficient political backing to make them a factor in bargaining. In considering the policy proposals that feed into the system, moreover, they often translate these into competing ideologies. Policy proposals can, however, also be studied in isolation from the policy-making process: in terms of the evidence on which their arguments are based and the logic of those arguments. The choice between Keynesian, monetarist and socialist economic policies is not simply the result of a play of power, nor is it entirely the result of decision-makers buying whichever theory promises solutions that fit their own goals. Some rational debate also takes place. In practice, one must admit, voters and politicians are too often influenced by the slogans associated with rival theories: 'short sharp shock' in penology, for example. But democracies are based on the notion that people compare the policies on offer, in party programmes for example,

Introduction

and it does the public little service if competing
options are treated as no more than political phenomena, rather than arguments in their own right. If
'to govern is to choose', then policy studies should
help make choice more sensible.

The slightly one-sided approach to the subject
referred to above is reflected in comparative studies, which tend to explain policy variations in
terms of environmental factors, notably political
forces. In some cases, however, different solutions
to similar problems may have been adopted because
policy-makers have reasoned differently. One can
not, of course, isolate decision-makers from the
political culture in which they operate or the political pressures to which they are subject: policy
choices are not laboratory decisions. Political
scientists are mainly interested in explaining why
different policies came to be adopted in different
countries, and will thus look at factors peculiar to
the country in question (its history, its politics,
its administrative system); or they will explain
similarities by the overriding force of common
economic factors. As suggested above, some people
may be more interested in clues to which policy
solves a particular problem most effectively: in
comparative policy studies as a way of assessing
which policy prescriptions appear to work out best
in practice. That said, other things are rarely
equal enough for cross-national comparisons of pol-
icy-effects to be treated as if they were laboratory
surrogates.

In most important fields of policy, the policy-
making process is obviously affected by a host of
environmental factors. The accumulated experience of
the past sets the scene. Historical traditions,
value systems and behaviour patterns (political
culture) tint the spectacles through which problems
are seen and define the sort of options that are
acceptable. Governmental arrangements affect the
policy-making process, defining how issues are
handled and where power is located. Equally important, solutions must in some way relate to existing
patterns of government. Just as the countries of
Europe have different political cultures, so they
have different institutional models on which to draw
and different legal systems into which they must fit.
If similar problems are differently handled, in other
words, it will not just be because of contingent
political factors but because countries have different concepts of the state and different traditions of
state intervention.

Introduction

The countries of Western Europe nevertheless
have many similarities. They have advanced, complex
economies, industrial and post-industrial, with considerable mixed-economy elements; urbanised societies with relatively high living standards; extensive
social services and educational provisions; liberal-
democratic political values, parliamentary institutions, free elections and competitive party systems;
extensive interest-group participation in policy-
making; sophisticated administrative systems, diver-
sified functionally and geographically, subject to
the rule of law and other controls; large professional bureaucracies headed by elected politicians.
In world perspective they form an obvious cluster.
   They shared a post-war era of economic growth in
which living standards seemed set to rise continually; in which, politically, everything pointed in
the direction of pragmatism and consensus, middle-
way policies that accommodated major interests
(something, much of the time, for everyone); in
which society appeared stable and challenges to the
established order were limited to a handful of
terrorists. Now they have similar problems, too, in
apparent reversal of earlier, happier decades. Since
the 1970s they face an international recession and
relative economic stagnation: the 'end of growth'
and the limits on government spending this brings.
Sectional interests nevertheless continue to make
claims on the state as a result of the 'revolution
of rising expectations'. Governments find it harder
to share out available resources and come under
sharper pressures as a result. They have to deal
with problems of unemployment and industrial restructuring whose origins often lie in international
developments outside their control. The 'end of
ideology' appears to have given way to a resurgence
of ideological politics, sharpening political conflict. The authority of the state is challenged as
interests pursue their claims through direct action
rather than the channels of representative democracy.
New claims, those of the environmentalists, for
example, or nationalist movements, add to the challenges governments face.
   All this should facilitate comparisons across
the states of Western Europe. In so far as comparison involves no more than the juxtaposition of
accounts, similar problem-field headings, this is
obviously true. It becomes more difficult if one
tries to go further. Social and economic problems,
though often similar, are rarely identical, because

Introduction

socio-economic structures are nowhere the same. The problems themselves may be differently perceived for reasons grounded in history. The interests that need to be accommodated vary in strength: power is differently distributed. While such accounts tell us why countries x, y and z have adopted the policies they have, and how they work out there, it is harder to decide whether a particular policy is the best solution to a common problem: whether countries x and y can copy z and expect the same results.

This is frustrating. Comparative studies nevertheless have a practical value. At the simplest level, though not a level to be despised, one can look at foreign experience to discover new ways of dealing with domestic issues that might not otherwise have been seriously considered by policy-makers. Although politicians and administrators, as well as policy advocates outside government, now have considerable experience of Europe through travel and reading, outlooks often remain parochial, dominated by national traditions and national experience. Comparative studies are a window on other styles of government and other ways of thinking about policy. The mode of comparison that involves no more than a juxtaposition of country-by-country accounts may thus stimulate thought about alternative ways of handling problems that might not have occurred to the policy-makers or have been dismissed by them as the advice of impractical theorists.

One can go further, however, without entering the methodological minefield of 'scientific' comparisons. Political scientists are sometimes too concerned with problems of methodology. Like policy-makers themselves, so political scientists, if they are to serve a useful role in society, must learn to content themselves with 'satisficing' rather than 'maximising' the empirical and theoretical underpinning of their prescriptions. A study of how - and how successfully - other countries have tried to solve problems broadly similar to one's own is bound to have some value, even if the lessons learnt are only suggestive, subject to a hundred methodological caveats by the scrupulous policy analyst. However rough-and-ready the comparisons, the conclusions of those willing to engage in the exercise will be as useful as those offered by the practitioners of other disciplines. None of the disciplines that deal with human affairs are capable of handling all the variables that influence behaviour and determine the outcome of government interventions. Economics, to

take an obvious example, seems hardly further advanced than meteorology as an applied science. Though
students of policy must remember the limitations of
their enterprise, they need not rank their contribution lower than that of others who offer advice to
policy-makers.

The countries of Western Europe share many problems at the present time: unemployment, inflation,
the need to restructure their economies, threatened
energy shortage, pollution, crime, the alienation of
'post-material' generations, unemployed youth, the
integration of immigrants, etc. The policy responses
of government are often similar too, especially to
economic problems. It even appears to some observers
that the common nature of problems leads to similarities in the policy-making process, notwithstanding differences in historical traditions, institutional arrangements and the balance of party-political
forces. Whether one emphasises the similarities or
differences between countries may well reflect no
more than the observer's starting point. To the
country expert, familiar with all the trees of his
forest, that forest may well look more different
from others than it really is. To those who take a
broader perspective, all forests may look decept-
ively similar. The editor of another volume on
policy-making asked: *"Could it be, then, that problems, policies and the process of formulating them
are converging in Western Europe?"* If the answer is
yes, one may be comforted by the fact that the un-
pleasant problems faced in Britain, and the unpala-
table solutions its government has adopted, are
shared by others. The less comforting conclusion is
that impersonal factors determine what governments
do. To accept that, however, is to undermine the
role of policy analysis. The important thing is not
to understand the world, but to change it.

The policy approach serves several purposes. It
is a way of studying government and politics, a
window on how the political system works. It is an
entry into policy debates proper, a way of acquiring
a broader understanding of the issues that face
society and the possible solutions to them. If,
finally, some governments seem to have coped better
with similar problems than others (on whatever definition of 'better' one's own political philosophy
dictates), it can make a practical contribution to
policy management. The scope of the present volume is
determined by the purpose for which it is primarily
intended: a textbook for students of European govern-

Introduction

ment and politics. For that reason, it has deliberately avoided a theoretical framework of its own
which would make it harder to fit courses organised
around other models. On the whole, it leaves the
drawing of cross-national generalisations of a theoretical sort to its readers. It should nevertheless
serve the three endeavours noted above.

Chapter Two

BRITAIN

H.M. Drucker & Richard Parry

Changing Policy Climates

Policy-making within the United Kingdom has
evolved through three different phases since the
second world war. After the national mobilisation
during the war, and the subsequent increased use of
state action by a Labour government, there was a
prolonged period of consolidation under Conservative
governments when the full-employment welfare state
was accepted by all substantial social and economic
forces; this lasted until the mid-sixties. From the
mid-sixties to the mid-eighties there has been a
period of alternation of party control of government, each in turn being discredited by the failure
of its economic policies and each searching for a
quick modernisation fix. From the late seventies,
however, even this idea was becoming discredited and
Britain embarked on a search for more radical solutions.

There is no agreement about what this more fundamental shift should be. There are those in the
Labour Party who would remedy the ails of Britain by
more thorough and more consistent use of state intervention. A second group, now dominant in the Conservative Party - in power since the 1979 general election - would apply the reverse strategy. They would,
to use one of their favourite metaphors, roll back
the state. A third school of thought hopes to rebuild around the old Keynesian and Beveridgian consensus, adapted by some tinkering with the machinery
of government; much SDP thinking takes this form.
This last is the least radical alternative on offer.

The war experience of a united, activist national government taught the coalition partners different lessons. Many Conservatives imagined that there
would be a going back to old ways with the cession

of hostilities, though the Conservatives who came to
dominate their party in the fifties soon accepted
the desirability of a national economic consensus.
Socialists saw the ability of the state to plan the
economy and remake the society; as they won the
first post-war election (1945) decisively, it was
they who earned the chance to implement their ideas.

The main features of the post-war settlement
were set out in the first two years of the Attlee
government. It was very important for subsequent
political argument that these features were established by a Labour government and (sometimes reluctantly) accepted by Conservatives. In the eyes of its
supporters and most economists, its greatest achievement was the attainment of 'full employment'. Mass
unemployment had been the greatest curse of working
people since the industrial revolution. Unemployment
reached historic peaks of 3 million in the thirties;
the ability of the first majority Labour government
to bring it down to negligible level would have been
a God-send even if it had been paid for with a high
level of inflation. In fact in the immediate post-
war period unemployment and inflation were low. Full
employment changed the relationship between capital
and labour: it gave organised labour a power in its
negotiations with capital which shattered many
ancient attitudes and institutions. With full employment came universal health and social security free
at point of use. The government had lifted great
weights from the backs of working people.

The post-war govenrment brought important parts
of industry into public and into national control.
Most of the power industries, public transport and,
latterly, steel, were nationalised. The government's
choice of industries to take over was partly dict-
ated by its supporters - Labour never nationalised
anything unless the relevant unions were in favour,
partly a matter of exhausted capital in the industry
and partly a question of the importance of the industry to the economy. There were those in the Labour
movement who saw themselves 'nationalising the
commanding heights of the economy'. Labour did not
attempt a revolution. Or rather, it had its own idea
of a 'British revolution'. It did not nationalise
capital: the government paid for the assets it took,
and it left the fundamental structure of the economy
in the hands of private capital. Arguably, it re-
freshed important sections of private capital by paying handsomely for assets (coal and railways in
particular) which had little profitable life in
front of them.

The high levels of taxation necessary to support the newly created social services and the enlarged public industry sector were burdensome to middle-class tax payers, but were accepted initially as part of the price of the post-war settlement which might so easily have been much less agreeable to the relatively affluent. The other change which followed almost unnoticed at first from the new settlement was a high level of public employment, about a quarter of the work force. Salaries for the enlarged public sector had to be found from taxation, but this burden was sweetened by the fact that many of the new jobs provided employment for the children of middle-class people.

The economy was run under the conventions of demand-management techniques which had been worked out for Britain by John Maynard Keynes. The government and the Bank of England set themselves the task of managing the economy as a whole, and not just the budget. This requires decisions by a few people in the Treasury and the Bank. In the short term, it does not necessitate building consent inside the civil service, parliament, the public or the trade unions. Indeed, the concentration of power within the British system suits Keynesian demand management techniques well.

Both major parties accepted features of a post-war consensus they would have preferred to avoid in a perfect world. For its part, Labour accepted that any further acts of nationalisation would be small and few. During much of this period it was not even committed to renationalising steel (the incoming Conservative government had unscrambled the still unfinished nationalisation reorganisation). Labour had learned from successive economic crises that further expansions of the public sector would have to be set in the context of the international valuation of the pound. Similarly, Labour governments would have to curtail their exuberant desire to expand the economy just by pumping more money into it: the first priority had to be exports, not domestic consumption. Labour had also decided during its first term of office to join the American-dominated North Atlantic Treaty Organisation: one consequence of this which she accepted, without enthusiasm to be sure, was a permanently high level of defence spending; the Korean War of 1950 stimulated the rearmament programme which tilted Labour's priorities from welfare to international security.

For their part, the Conservatives also had adjustments to make. They had to accept the reality

and permanence of nationalisation together with a
high and expensive level of publicly provided social
services. They had to accept the more prominent role
of the trade unions as an 'estate of the realm' and
to consult their leaders on economic policy. At the
same time, some once solidly Conservative institutions like the Church of England ceased to matter and
other reliable institutions like the universities
and the BBC became restive.

Of more sentimental importance was the loss of
Empire. Churchill could fume that he had not become
Her Majesty's First Minister to witness the end of
the Empire: that is precisely what British governments from 1945 did. The cost of colonialism was too
high, the American allies too contemptuous, the
benefits too ethereal for the project to be sustained. The failure of the invasion of Suez in 1956
showed that Britain could not pay the economic price
of colonial adventures against American opposition;
the decolonialisation of Africa which followed had
become inevitable. Both sides were willing to pay
these prices out of a combination of political calculation - the Conservatives, fearing a repeat of
1945, sought to show that they could manage the welfare state better than Labour - and economic interest - Labour could not discount notions of personal
gain in an age of rapid economic growth.

But the era was short-lived. The consensus began
to crack in the early sixties, sustained economic
growth came only in bursts of 'stop-go'. Britain had
begun to grow less quickly than other countries:
middle-class people began to return from summer holidays on the continent with the uncomfortable feeling that the Europeans were better off than they.
Working-class people started to go to Spain, rather
than Blackpool or Skegness, for their holidays and
suddenly Britain looked tatty. Harold Wilson sensed
the mood of the country in 1964 and again in 1966
when he promised to remake the nation in the heat of
the technological revolution.

When Wilson first became Prime Minister in 1964
his party had been out of office for thirteen years.
Not surprisingly, they wanted to make substantial
changes in the machinery of government. The old
machinery was too closely associated with the use
their enemies had made of it in the period since
Labour last ruled. But Wilson and his colleagues
launched a wave of reorganisation of the institutions of public administration which cannot be solely
accounted for in this way. The truth was that much
of the establishment, not just power-hungry social-

ists, thought that one of the chief explanations for
Britain's poor social and economic performance over
recent decades rested in outdated machinery of government.

'Efficiency' not 'equality' was the true hall-
mark of the Wilson regime. An attempt was made to
reform the civil service, specifically the recruitment, training and promotion of the service. Large
firms were encouraged to merge into even larger
firms. The social services were reformed and expanded where it was believed that change would produce
a more productive work force: but they were not
expanded universally, and charges for prescriptions
were introduced. An Anglo-French agreement was
entered into to build a supersonic passenger plane
just to beat the Americans to the punch.

In the middle of its period of office the Wilson
government also latched on to the notion that entry
into the European Communities would be the panacea
for the ails of British industry. This enthusiasm
was never entirely shared by the Labour Party; but
the potential for embarrassment which this gap
opened up was not realised as the French vetoed the
application.

But as the first Wilson administration faced the
growing economic problems of the country, it became
convinced that reform of the trade union movement
was the fundamental necessity and that it, as a
Labour government, had a unique chance and responsibility to carry through such a reform. A reformed
trade union movement, one in which power and authority rested in officials who could be held account-
able for their actions, could negotiate and fulfil
bargains as part of a national 'prices and incomes'
policy. In the waning months of its office the
government attempted to legislate for the reform of
the trade unions and failed. Shortly thereafter it
fell.

The first Wilson government (1964-70) was noticeable for the first widespread civil disobedience in
Britain - over its support for the United States in
Vietnam - to be seen since the war. It also witnessed the failure of law in Northern Ireland. In both
these cases, as well as in the government's defeat
by its trade union supporters, Britain began under
Wilson to witness a failure of government. The economic failures were still very much the centre of the
political stage, but the coincidence of the two
kinds of failure suggested deep difficulties in the
polity.

The Heath government, elected in 1970, was

publicly committed to a reversal of the consensus
which had guided policy since the war. This commitment (to the principles of 'Seldsdon Park', after
the hotel in which the ideas were hammered out) was
not followed through. Within two years of taking
office the government was having recourse to much
the same nostroms as its predecessors. Heath tried
to rejig the machinery of central administration;
he perfected systems to plan government expenditure
as an economic variable (PESC); to monitor the progress of long-standing government policies (PAR);
and, above all, he drove the nation into the European Communities. Like Wilson Mark I, he tried to save
the country by modernising the machinery of its
operation. Like his predecessor, he failed to win
reelection. Heath called a general election and
asked for support against the miners who were on
strike at the time; he wanted a mandate to confirm
his right to represent the national interest. But
such a divisive stance was rejected; Labour won more
seats than the Conservatives in the February 1973
election.

The second period of Wilson-led Labour government was, initially, a repeat of the first. Each
party in opposition had raised the rhetorical stakes
against its opponents. Each promised more in office.
Each spent its first months of office pursuing manifesto pledges without regard to the degeneration of
the economy. Under each successive government the
rate of unemployment reached unprecedented levels;
and each proved incapable of doing anything about
it. At the same time the rate of inflation rose. The
twin rise in inflation and unemployment had been
thought to be impossible by most Keynesian economists: economic policy-making was meant to be a
matter of deciding on what trade-off between the two
evils the country could afford.

By the summer of 1976 there was no longer anything approaching an establishment consensus on how
to proceed. Three lines were being canvassed. The
left of the Labour Party, with some support from
unreconstructed Keynesians, advocated high tariff
walls against increased imports, expansion of domestic demand to soak up demand and reduce unemployment, and a strict prices and incomes policy to
control inflation. With the exception of the tariff
policy (which was clearly inconsistent with Britain's
EC commitments) this was the old policy as before.
It was very much in line with Labour's historic role
of protecting the jobs of manual labour.

In office, faced with the currency crisis of
1976, Labour actually accepted the terms of a large
loan from the International Monetary Fund. The IMF
had been won over to deflationary economic policies
and demanded cuts in public expenditure and continuation of wages control as the price of its loan.
The government agreed. This was not a policy which
Labour could operate with any comfort or skill, and
it remained for the Conservative government elected
in 1979 to seek to justify the newer stringency.
Conservative solutions to Britain's government crisis involve a weakening of all the institutions which
had been accepted as part of the post-war settlement:
the unions, the public administration of welfare;
the high level of personal taxation, the large public sector and now, generally, the notion that if
things go wrong the state will provide a safety net.
Ideologically, the Conservative plan is reactionary.
It is based on a wish to return to Victorian simpli-
cities. Its believers have dominated the offices of
government since the 1979 election but they have
failed to acquire the patina of success and authority for their plans.
     To the disgust of the government, many civil
servants, the majority of the serious newspapers and
much of the public administration industry remain
committed to a form of the old post-war consensus.
This group would like to reorganise the trade unions,
that great unfinished piece of 1970s business, and
curb their political influence over the Labour
Party; and they would like some form of proportional
representation to replace the present electoral
system. The latter reform is very much in their
interests as an organisation, but it should not pass
unnoticed that such electoral reform would most
likely mean that the real negotiations about changes
in government and policy took place between MPs
behind the scenes after elections. It is a recipe
for no change. It is this constituency which the
newly formed Social Democratic Party and its Liberal
allies are hoping to win over.

Economic Policy

     The three responses of the political parties -
Labour state intervention, Conservative rolling back
the state, and Liberal/Social Democratic incremental
management - are reflected above all in economic
policy, the central arena of party competition and
of preoccupation by political elites. The two world
wars, especially the second, were the crucial periods

of expansion in British state activity. Britain
emerged after 1945 with high public expenditure,
public employment and taxation, and also with a consensual approach to economic management based on
Keynesian ideas of regulating aggregate demand. The
tools seemed to be in place for sustained economic
growth that would avoid the mistakes of the 1930s
and carry through the wartime spirit of national
solidarity into peacetime tasks. In the event, many
of these hopes were not fulfilled, and Britain's
alleged economic failure became a centrepiece of
political debate.

Analysis of economic policy can best be approach-
ed through a review of the symptoms of poor performance, some of the causes and explanations that have
been advanced for them, and the remedies proposed by
the parties. Ultimately, postwar economic policy
reflects a dilemma: is there a way out for the
British economy, a route to prosperity given the
right policy instruments, or is the long-run British
weakness in manufacturing industry, evident since
the turn of the century, proof that Britain faces
relegation to the second division of industrial
economies? Increasingly, pursuit of economic breakthrough has been overtaken by a dull acceptance of
decline which diminishes the responsibility of politicians to correct it.

Symptoms

The symptoms of Britain's poor economic performance are not in doubt. Britain has slipped steadily
down the 'league table' of GNP per head. In 1945 she
was one of the most prosperous nations in the world;
by 1960 she had been overtaken by France, Germany
and the other Northern European countries; by 1983
she was well behind all the other members of the
European Community except Italy, Ireland and Greece,
and seemed to have little potential to improve her
position by the end of the century.

There is a political paradox here, in that the
30 years after 1945 saw the steadiest period of
economic expansion in modern British history, with
an annual increase of real GNP of about 3 per cent.
Britain's cyclical pattern of growth was much smoo-
ther than that of many other countries, and in no
year from 1952 to 1974 did real GNP contract. The
problem was simply that Britain's GNP only doubled
in real terms during this period, when it trebled in
France and Germany. Because Britain was growing
richer, the extent to which it was being overtaken

was masked. The steady growth of the 1950s could be
represented, and often was in political debate in
the 1960s, as an unsatisfactory base line which
could readily be improved by the right policies. But
the opposite happened; even what Britain had was
taken away from her. Since the recession of 1973-75,
generated by increased oil prices, growth has been
less than 1 per cent a year, while the recession of
1979-81 saw a decline of 5 per cent in real GNP and
of 18 per cent in the output of manufacturing industry.

A parallel indicator is that of low productivity.
Output per head was comparable to that in France and
Germany in the 1950s, but has increased at less than
half their rate. Capital investment in British fact-
ories seems to have yielded lower output than comparable investment elsewhere. In multi-national
companies with similar processes - like motor manu-
facturers - British productivity levels have in some
cases been only half of those found abroad.

A consequence of low economic growth is low
levels of employment generation. The British workforce has increased significantly in the post-war
period, from 23 to 27 million, though this is less
than the increase in most industrialised countries.
The number of women rose significantly, the proportion of married women of working age in jobs more
than doubling. But the private sector has, over the
whole period, seen no net increase in jobs at all;
an increase of over 2 million by the mid-1960s was
entirely lost by the early 1980s. In the public
sector, on the other hand, employment rose from 6.3
million in 1951 to 7.6 million in 1981, over 30 per
cent of the employed workforce.

The result of the private sector's failure to
create net new jobs - as increases in service occup-
ations failed to match a decline in manufacturing
employment which became precipitous in the early
1980s - was a vulnerability to high unemployment.
For long this was averted by public sector growth,
by a demographic pattern which caused a reduction in
the working age population in the late 1960s, and by
the tendency of older men to retire from work earlier. Until the 1970s unemployment rates of more than
2 per cent were rare. But in 1972 unemployment passed 1 million for the first time in the post-war
years, and since 1975 it has never been below that
figure. In the late 1970s it stabilised at about 1.5
million, but from 1980 to 1982 it rose rapidly to
over 3 million, or 12 per cent of the workforce.
Public sector employment creation ceased, and the

demographic bulge of the 1960s' baby boom came on to
the labour market. Long-term unemployment became a
major problem, with one-third of the total out of
work for more than a year; many school-leavers had
no prospect of a job. Britain now had an economic
ill it had previously been spared by demographic
good fortune and full employment policies - a recession translated directly into job-shedding.

A further major symptom of economic weakness is
inflation. Unlike unemployment, this was perceived
as a 'British disease' even when inflation was low
in the 1950s. Prices rose every year (except briefly
in 1959) in a cost-pull pattern fuelled by the pursuit of money wages. This tendency was reinforced by
the British practice of centralised wage agreements
supplemented by local 'wage drift'. Low productivity
growth and weak management aggravated the inflationary tendencies, and national wage norms injected
further distortions. As attempts to control inflation became more explicit in the 1960s and 1970s,
the results were if anything worse. 10 per cent
annual inflation became normal in the early 1970s,
and wage explosions in 1974-76 and 1979-80 took it
temporarily to over 20 per cent; the record 26 per
cent was in mid-1975. For long, compulsory wage and
price restraint seemed the only means of restraining
inflation, and in the process many voters developed
a vested interest in high inflation as a means of
reducing the real cost of loans (especially on house
purchases).

For over 30 years Britain's balance of payments
was a perennial source of concern. Britain's share
of world trade halved between the mid-50s and the
1970s, and there was a surplus on visible trade in
only one year (1971) in the 1960s and 1970s. Although the deficit was often covered by invisible exports, all too often deflation of domestic demand in
order to reduce imports seemed to be the only policy
response. The balance of payments ceased to be a
prime issue because of the windfall discovery of oil
and gas in the British continental shelf in the
North Sea. Oil started to be produced in commercial
quantities in 1976; by 1980 Britain was a net exporter. Production is expected to reach its peak in
1985 but will continue at a substantial rate until
the end of the century. By 1983, oil and gas were
contributing nearly 5 per cent of GNP and 6 per cent
of tax revneues, and were boosting the balance of
payments - now barely in surplus - by over 10 billion
pounds compared with the mid-70s.

In political debate in the 1970s it was frequent-

ly suggested that the government in office during
the peak years of oil production would derive an
impregnable political position. Indeed, oil has
merely moderated some of the worst effects of the
recession and has threatened Britain with the
'Dutch disease' - the phenomenon observed in the
Netherlands and Canada of natural resources leading
to an overvalued currency, excessive current consumption and inadequate investment in other industries.
Far from solving Britain's economic problems, oil
has highlighted the intractable state of the rest of
the economy and the symptoms that threaten to remain
long after North Sea fields are exhausted and
Britain has to earn currency to import its energy.

Explanations

The combination of low growth, sluggish productivity, poor employment generation and constant inflationary pressure could never be an acceptable one
for the British political system. British history
and national achievement gave her high, not low,
pretensions to economic success. The sharp differ-
entiation in the party system between Labour and
Conservative made it electorally imperative for each
to offer an alternative to relative economic decline.
Correspondingly, there was a need to establish an
explanatory framework for Britain's performance that
could serve as a guide to policy.
Perhaps the most influential explanation was the
one that saw excessive concentrations of power as
the root of the problem. This has usually been expressed in terms of hostility to the trade unions,
which organise over half of the workforce, a high
proportion by world standards. Labour disputes were
so often the visible face of economic difficulty
that a consensus grew up in the 1960s that 'restrictive practices' at the workplace were the brake on
productivity and exports. Equally relevant to this
explanation, but frequently neglected, is the traditionally weak policy on competition and mergers,
and the fact that cartels or monopolies were often
tolerated by public policy. The Monopolies Commission, a government agency, lacked teeth, and for a
time in the 1960s amalgamations of major firms were
encouraged by government. Employers were able to
shelter behind the poor public image of unions and
their leaders.
If trade union practices are the 'treason of the
workers', a second explanation - the orientiation of
British capital to financial manipulation rather than

23

Britain

industrial growth - is an alternative 'treason of
the employers'. Since the nineteenth century Britain
has had an exceptionally open and internationally-
aware business structure, established by her trading
tradition and world empire. Freedom to invest abroad
and repatriate profits, and protection of a sterling
exchange rate intended as a tool of international
transactions, have become central assumptions of the
political economy. Britain also has an active stock
market, with much investment dependent on raising
equity, and thus on business confidence, rather than
through bank loans or internally generated profits.
Dependence on United States loans after the second
world war prevented any alternative strategy and
made Britain vulnerable to the withdrawal of foreign
capital and attendant balance of payments crises.
The result was a tendency to over-value the currency
and resist parity changes that might benefit domestic
industry. The maintenance of the pound at $2.80 from
1949 to 1967, before devaluation to $2.40, was
defended on this basis; that constraint was removed
by the decision to float the pound in 1972, but
subsequent exchange rate movements were erratic and
owed too much to intangible international confidence
at the expense of industrial competitiveness. The
exchange rate plunged in 1976 and prompted domestic
deflation under an International Monetary Fund package; the rate soared in 1979-80 and crippled export
prices.
     Implicit in this explanation is the idea of a
pay-off from government to private capital, especially under the Conservatives. Relaxation of credit
control in 1971 was a major boost to speculative
capital, prompting a property boom in the mid-1970s,
the subsequent collapse of a number of major firms
and a 'lifeboat' operation by the Bank of England.
The abolition of exchange controls on overseas
investment in 1979 gave a major new freedom to mobile capital. Labour governments have also been
accused of excessive deference to private capital,
personified in the 'gnomes of Zurich', the IMF
review teams and the governors of the Bank of
England. The deflationary packages of 1966 and 1976
were seen as unnecessary national humiliations by
the Labour left. The support of Labour governments
for joining the European Community also fits into
this pattern, because the idea of the common market
with free movement of labour and capital can be
represented as the loss of national political corr-
ectives against the actions of 'unpatriotic' financiers.

Britain

A third sort of explanation is that the British
public sector is too big - that excessive public
expenditure, taxation and borrowing have preempted
resources and 'crowded out' the private sector. Such
a model has proved attractive to many Conservative
politicians because it suggests a dichotomy between
a productive (private) economy and a parasitic public sector. There are problems with this theory.
Conservative governments accepted almost all of the
post-war industrial nationalisations, recognising
that some essential utilities could not be viable
outside the public sector. The public share of national produce (46 per cent in 1983) is comparable to
that in other western European countries and public
sector deficits have been smaller than most. There
is little evidence that the crowding-out effect has
reduced the available amount of investment capital;
more important has been the lack of suitable projects. The thesis that high marginal rates of personal taxation are a disincentive to entrepreneurial
effort has not been substantiated by experience
since these rates were reduced in 1979.

Equally widely held is the belief that lack of
investment is a major cause of Britain's economic
weakness. This was reflected in the ambitious investment plans of the public sector in the 1960s and
the grants and loans available for private sector
investment in favoured locations. But the problem
seems to be less a shortfall in the supply of funds
than high interest rates and the long-run decline in
the real rate of return on capital. Investment is of
little use if it is in the wrong kind of industry
and fails to yield efficiently produced and saleable
output. Nationalised industry investment has had
poor results, and one initiative - in steel in the
early 1970s - led to disastrous over-capacity. In
the 1950s Britain was the world leader in atomic
power and civilian aircraft and the European leader
in motor manufacture. This lead was eroded more by
misguided investment decisions made by both public
and private sectors than by lack of investment, so
that by the 1980s imported technology and equipment
were the rule in all these sectors.

.A final most over-arching and least precise
explanation is that Britain suffers from a 'non-
industrial spirit' and is not interested in the
highest economic performance. Surveys indicate that
British workers value leisure and the freedom to
control work practices, and decline to give all to
their employers or company. British elites are also
alleged to have value systems orientated to cultural,

pastoral and social pursuits. Compared with the
civil service and the professions, British management appears to have failed to attract high-quality
recruits. It is often more profitable and gratifying
to provide services than engage in the direct production of goods. This concentration is not necessarily bad, as service industries can export what
they produce and provide buoyancy during a recession; but the basic manufacturing processes have not
matched the level of achievement found in the 'post-
industrial' sector.

Remedies

The political response to these explanations has
lacked consistent direction. Alternation between
governments, and the contrast between their policies,
has increased. It has been described as a 'Jekyll
and Hyde syndrome' in which parties promise ambitious and ideologically-motivated policies while in
opposition, take precipitate initiatives when coming
into office, slowly adapt to the 'real world', and
then lose a general election to their opponents who
embark on similarly unrealistic plans. This kind of
argument is common among the liberal, incrementalist
centre, but is itself ideologically loaded and con-
ceals the continuity of underlying policy between
governments. But it does provide a clue to the contrast between the tensions and failures of the 1960s
and 1970s and the early post-war years which are now
taking on a nostalgic glow.

The decade after 1945 was dominated by resource
shortages. Rationing of food continued, and growth
ran into balance of payments constraints. In 1949 a
large devaluation of the pound recognised the relative decline of the British economy against the
American, but internally the record was better. The
trade unions were acquiescent in wage restraint,
inflation and unemployment were notably low, demobil-
isation was completed and private-sector growth was
generated. The Conservative election victory of 1951
was a tactical disaster for an exhausted Labour
government, but in policy terms it was a natural
progression, as the need was for a liberal economic
policy to promote export-led growth. In many respects the 1951-64 government was successful in
achieving this. The effortlessness of economic
growth became a commonplace in political debate, as
in Anthony Crosland's *The Future of Socialism*. The
dominant policy theme was that of 'fine-tuning' -
the regulation of the economy to steer the narrow

path between inflation and deflation. By using marginal changes in taxation and credit control, periods of excessive demand or of higher unemployment could be ridden out. Failure to steer this course was liable to be punished by the electorate in what was the golden age of post-war two-party politics. The Conservatives were successful in delivering election-winning budgets in 1955 and 1956; when in the latter year Prime Minister Harold Macmillan said that 'most of us have never had it so good', he was accurately reflecting the success of economic management.

Still, there remained a grumbling dissatisfaction about Britain's economic performance, all the more pointed as her world role contracted. Despite the consensus nature of many policies, Britain never acquired the equivalent of Germany's 'social market economy' or France's 'planification' - the idea of a sound, stable, policy frame within which economic performance could be maximised. Real achievements were interpreted as only provisional successes upon which a major economic upsurge could be imposed; notions of real limitations on Britain's economic potential were derided.

The policy themes that were to dominate economic debate for nearly twenty years began to emerge in the early 1960s. They included pay policy - the desire to impose an economy-wide limit on wage claims; planning and regional development that would direct industrial growth to correct regional imbalances; joining the European Community in order to extend the market for British goods and provide a competitive spur to home industry; higher public expenditure, especially on capital investment and education; and trade union reform - the contention that only institutional changes could release the full potential of labour.

All these approaches had a common characteristic: they were 'hands-on' policies, 'purposive' in the language of the time, with governments actively seeking to assume responsibility for the state of the economy and credit for policy remedies. In the 1960s and 1970s this hands-on spirit overcame the alternative hands-off approach - normal until the 1940s - in which government sought to distance itself from the performance of the economy and wished to control only a narrow range of economic variables. The 1970 Conservative government of Edward Heath came to office intending to return to the old tradition, but soon reverted to the kind of policies associated with Harold Wilson as Prime Minister from 1964-70 and 1974-76. It was only with the 1979 Conservative government of Margaret Thatcher that a full-blown hands-off policy was pursued.

The hands-on approach owed much to electoral needs. With the development of more sophisticated political marketing and television coverage, and of informed public debate on economic matters, parties wanted to offer a novel and inspiring prospectus. Harold Wilson's call for a 'white-hot technological revolution' in 1964 was thin on substance and damaging in the long-run, but a politically brilliant invocation of dynamism and of potential in the economy waiting to be aroused. Heath's 'new style of government' and 'Selsdon man' philosophy were different but similarly bold in their promises and their wish to grapple with the problems. Wilson and Heath had both been members of the war-time civil service and were attracted to the spirit of national mobilisation and solidarity found in those years. In Wilson's case, the fighting spirit of Dunkirk was invoked to defend the $2.80 parity of the pound, which was maintained for a damagingly long period before devaluation became inevitable in 1967. In Heath's case, entry into 'Europe' served as the symbolic core of his government's strategy. In each government, senior civil servants were influential in policy-making, and their perspectives tended to dominate over those of party activists.

Pay policy followed a remarkably similar course in the 1964 and 1970 governments. First tried under Macmillan with the 'pay pause' of 1961, policy soon became much more comprehensive and rigorous. A cycle developed of pay freezes (in 1966, 1972 and, with a norm well below inflation, in 1975); re-emergence with 'severe restraint' and a periodical easing of the policy (in 1967-69, 1973-74 and 1976-78); and a free-for-all catching-up as powerful unions sought to restore certain differential while claiming comparability with other occupational groups. The whole cycle had short-term success, but ultimately it fuelled inflation because the catching-up awards tended to run well ahead of prices and resulted in real gains to those in work. Both Conservative and Labour parties were in principle opposed to statutory wage constraint, but in power were curiously attracted to it. It remains the classical social democratic policy, and was initially supported by the SDP.

Regional policy similarly became prominent in the early 1960s as an attempt to correct the divergence in regional unemployment rates, a persistent problem in the 1950s. Industry was channelled to the

North of England, Scotland and Wales through investment grants, restrictions on development in the
South East and Midlands, and ultimately labour subsidies (the Regional Employment Premium). In the
wider field of economic planning, 'tripartism'
between government, employers and unions was developed through the National Economic Development
Council (1962), the 'declaration of intent' on wages,
prices and productivity (1964) and the National Plan
(1965). The latter was prepared by the Department of
Economic Affairs, established in 1964 in an attempt
to remove what was regarded as the 'dead hand' of
the Treasury on the promotion of growth. The grandiose targets of the National Plan were overtaken
within months by economic circumstances, but more
modest versions continued in the work of the NEDC,
which was especially influential in the Callaghan
government of 1976-79. What these instruments lacked
was the means of relating the goals of government to
the operation of the real economy. Regional policy
shifted some jobs around the country and did aid the
convergence of regional unemployment rates; but it
discriminated against many parts of England, and by
1981 the West Midlands had a higher unemployment
rate than Scotland. Political calculation (the wish
to protect Labour party strength in the peripheral
parts of Britain, or to correct Conservatrive weakness there) became an element of regional planning,
but it could not compensate for the lack of growth
in the economy as a whole. Corporate consultation in
bodies like the NEDC was congenial to government but
failed to secure performance at the factory level.
    Policy towards the European Community occupied a
dominant and probably excessive place in political
debate in the 1960s and 1970s. Britain had refused
to become involved in early moves towards European
unity (the Coal and Steel Community and the Defence
Community) out of a mistaken belief that she was
operating in a wider geopolitical sphere. Britain
lost the chance to influence the structure of the
European Economic Community established in 1958, and
so could not prevent its expenditure from becoming
dominated by the Common Agricultural Policy. When
Britain sought to join in 1961, acceptance of this
structure became an entry-price for free trade within Western Europe and British influence in European
affairs; Britain switched from a position of haughty
detachment to the equally undignified one of import-
unate suitor. EEC entry, though opposed by a 'country' coalition of Labour economic nationalists and
Conservative imperialists, was the favoured policy

29

of the 'court' of diplomats and financiers; it became irresistible to Macmillan, Wilson and Heath.
Applications in 1961 and 1967 were vetoed by de
Gaulle, who perceived the ambivalence in Britain's
position, but entry was finally secured after pro-
tracted negotiations in 1973. A referendum on continued membership passed comfortably in 1975, having
been used by Harold Wilson as a tactical device to
obtain Labour Party acquiescence to entry. Thereafter, results of membership were disappointing. The
oil crisis of 1973 had ended the Community's period
of rapid economic growth; net budgetary contributions became increasingly onerous; and free trade
within Western Europe seemed to encourage imports
rather than exports. In 1979 Britain declined to
join the new European Monetary System and became
involved in annual wrangling to reduce its inequit-
able share of the Community's budget. The main
British benefits of membership accrued to farmers
and to workers where American and Japanese investment was attracted by access to the European market.
EEC membership may have been politically necessary,
but it quickly became clear that it had never been
the variable making the difference between economic
success and failure for Britain.

Another remedy available in the 1960s was an
increase in public expenditure. Between 1960 and
1975 it rose from 41 to 49 per cent of national
product, up by 80 per cent in real terms. There was
major capital investment in roads, schools and hospitals, and public employees were hired to staff the
new facilities. The aim was to provide what were
conceived as the preconditions of economic performance - an educated, well-housed and mobile workforce
and a range of government supports for industry. In
1961 a system of forward-looks on public expenditure
in constant price terms (the PESC system) was dev-
ised, and it came into full operation at the end of
the 1960s. This had many virtues as a tool of economic management by relating public expenditure to
the consumption of resources in the economy as a
whole. But it encouraged governments to plan the
growth of public expenditure on the basis of unrealistic projections of economic growth, and it proved
particularly disastrous in controlling expenditure
in times of high inflation, as it operated on the
volume of services rather than their cost. By the
mid-1970s public expenditure was widely being described as out of control, with final out-turns way
ahead of initial plans and rapid growth in the relative price effect (the tendency of public sector

30

prices for wages and services to run ahead of the
rest of the economy). Finally. the PESC concept had
to be severely compromised by the introduction in
1976 of cash limits in current money terms on most
items of expenditure. Real cuts in public expenditure were enforced in 1977 and 1978 as part of the
conditions for the IMF loan; capital expenditure
suffered particularly.

The final remedy that failed to work was trade
union reform. Pressure for this arose from the widespread identification of the trade unions as the
'devil in the machine' that was impeding growth.
Britain's wage bargaining system was different from
that of most European countries; rather than occasional major disputes about wage contracts, there was
a continuing series of smaller strikes, often related to wage differentials between occupational
groups and inter-union disputes about work practices.
Occasional strikes in major utilities with serious
effects on the public (train drivers in 1955, seamen
in 1966, coal miners in 1972 and 1974) could also be
interpreted as a challenge to government. In 1968
the report of the Donovan Royal Commission on Trades
Union represented the culmination of a gradualist
approach, with its concentration on building up
sound bargaining procedures and trade union negot-
iating rights. Thereafter, governments went for a
quick and comprehensive fix; but they did not find it
easy. The Wilson government's proposals for cooling-
off periods and strike ballots in 'In Place of
Strife' in 1969 were withdrawn under pressure from
the Labour Party, with the TUC making a 'solemn and
binding' undertaking to put its house in order on
inter-union disputes. In 1971 the Heath government
passed the Industrial Relations Act, a complex legal
structure worked out in opposition, which tried to
create the concept of a registered trade union which
would receive bargaining rights in return for abst-
aining from 'unfair industrial practices' as policed
by a National Industrial Relations Court. The tactical coup of the TUC in persuading unions not to
register and securing a pledge of outright repeal of
the Act from the Labour opposition prevented the
policy from working, and it turned into farce in
1972 when a few trade-unionists succeeded in their
aim of being sent to jail for contempt of court, and
when the only strike ballot (on the railways) showed
a clear majority for the union.

As a reflex, the 1974 Labour government was
exceptionally kind to the unions, giving them

31

unheard-of corporate rights and patronage. The dream
of 'doing something' about the unions receded. The
Thatcher government's legislative measures (the
Employment Acts of 1980 and 1982) were carefully
framed to give employers the right to seek court
injunctions against certain types of industrial
action; the government was removed from the firing-
line. In the end, it was the recession that curbed
the unions' influence and membership strength, not
the legislative structures sought by successive governments.

Public confidence in, and government pursuit of,
these remedies of the 1960s and 1970s came to an end
with the 1979 general election. The most dramatic
circumstance of post-war economic policy is the way
that the Thatcher government survived far worse
economic indicators than any of its predecessors, to
emerge with an election victory in 1983. Part of
this victory derived from the Falklands campaign of
1982, the equivalent distracting foreign adventure
to Suez in 1956 or Rhodesia in 1965, though with a
positive outcome for the government. But even before
then the opposition were having difficulty making
political capital out of the 5 per cent fall in GDP,
13 per cent unemployment, a peak inflation rate of
22 per cent, and a major loss of competitiveness by
British industry. The Thatcher government had some
success in turning the previous strategy on its
head: instead of loading responsibility on to its
shoulders, it emphasised its separation from the
real economy, inviting business and the labour force
to bear their own responsibility for its economic
performance.

The roots of this success lie in the traumas of
the 1970s. After its doomed fight against devaluation, the Wilson government achieved some stability
in 1969-70, with a balance of payments surplus and
an excess of taxation over public expenditure. A
more expansionary budget in 1970 might have secured
a Labour victory, but the new Conservative government launched its 'Selsdon man' philosophy, only to
abandon it in 1972 in favour of stimulating growth.
This proved over-ambitious; the oil crisis of late
1973 coincided with a breakdown of pay policy, major
public expenditure cuts, a miners' strike causing a
three-day working week for industry, and Conservative
defeat in the general election of February 1974.

Labour returned to office unprepared and its
policies were perverse ones in a recession; by 1975
it, too, had to retrench sharply. Although unemployment stuck above 1 million, the Callaghan government

32

in 1977-79 was able to relaunch modest economic
growth and even talked of an 'economic miracle',
soon to be underpinned by North Sea Oil. But an over
-ambitious pay policy precipitated the 'winter of
discontent' of 1978-79, with disruptive public
sector strikes; election defeat soon followed. And
so each party had seen the ignominious collapse of
its government in the depths of winter, to the
alienation of its own supporters. Public confidence
in the policies they had pursued never recovered.

The Conservatives came to office in 1979 committed to monetarism, the converse of Keynesianism;
instead of controlling demand in the economy, it
concentrated on the supply of money and the balancing of the budget. The intellectual attraction of
monetarism was, first, that it was different and
untried and, second, that it appeared to fit the
experience of the 1970s: the credit boom of 1972-73
had been followed by the high inflation of 1974-75.
Quickly, monetarism itself failed. In 1980-81, the
monetary growth target of 7-11 per cent turned into
an outcome of 20 per cent. The government soon
abandoned its attempt to keep money supply growth
on target, even at the price of excessively high
interest rates and exchange rate. Moreover, the
Conservatives had the embarrassment of seing public
expenditure in real terms and as a proportion of
GDP increase in each year of their first term as a
consequence of the recession.

Monetarism was replaced by the 'medium-term
financial strategy' based on reducing the public
sector deficit. There was no pay policy, though
public sector wages were squeezed through cash limits; nationalised industries, especially steel, were
allowed to contract, but there were only two major
denationalisations (aerospace and road freight) in
the first term; public expenditure planning moved
from volume to cash terms. The response to unemployment was in the form of manpower schemes to keep
some workers, especially the young and old, off
unemployment registers. Many workers failed to follow the lead of their unions in fighting government
policies on closures; the dispute in the engineering
industry of 1979 and the national steel strike of
1980 may have been the last of the old breed. Significantly, real take-home pay has been maintained, and
inflation fell faster than expected in 1982 to below
5 per cent; for those in work, the economic climate
has been benign, and there was even a minor consumer
boom, symbolised by the video recorder - 25 per cent
of British homes have one, the highest proportion of

any country.

The Conservatives have provided a psychological equivalent of the defeat in war and the destruction of economic and social assumptions faced by France, Germany and Japan. It is no accident that the humiliation of these countries came to be regarded as a spur to action. The new approach represents a radical alternative to the Callaghan government's mood of protection against weakness, described as 'steady as she sinks'. This was the echo of 1960s policies and is to some degree still evident in the 'extreme centre' spirit of Social Democrats and Liberals. Unlike Heath, Thatcher challenged the assumption that the electorate would throw out any government that failed to stimulate demand during a recession and, as much through style as substance, convinced the electorate that a temporary deterioration of the economy was a necessary catharsis.

The third strategy - thoroughgoing state intervention and economic nationalism, associated with the Labour left wing - has never really been tried, because the Attlee, Wilson and Callaghan governments sought a world role in economic and political alignment with the United States. Such a strategy, involving tariff barriers and the control of investment decisions, would accept that Britain could not compete on world markets and seek to mobilise productive resources on a social rather than an economic nexus. As such, it would deny the historical openness of the British economy. But the Thatcher strategy was radical and once seemed unthinkable; if it fails, Labour's policy may be the only option left.

Social Policy

The post-war welfare state is one of the most celebrated and stable features of British public policy. It stems from the wish of all parties in the 1940s not to return to the wasteful and undignified social policy of the 1930s, symbolised by the means test, but to sustain the health and welfare of the population. The welfare state was based on two wartime initiatives. The Beveridge report of 1942, written by former civil servant. Sir William Beveridge, called in trenchant terms for a comprehensive scheme of social insurance against unemployment and sickness and of retirement pensions, to be 'earned' by contributions, backed up by family allowances and full employment policies. The 1944 Education Act, steered through by a Conservative minister, R.A. Butler, ensured free secondary education for all and

expressed a faith in the power of meritocratic education which was extended to higher education in the
1960s.
    These were measures with cross-party support, but
the implementation of the policies by the Attlee
government ensured that the British welfare state
was positioned on the left wing of the political consensus. This was particularly true of the National
Health Service, founded in 1948, which took nearly
all hospitals and their staff under direct state
control and was free to patients on a non-contribut-
ory basis. The system of retirement pensions was also
actuarially more generous than Beveridge had intended. These policies were accepted, with some reluctance, by the 1951 Conservative government, and for
nearly thirty years British social policy was remarkably consensual. Party conflict was about who could
spend more on public housing, education, hospitals
and pensions, and few politicians were prepared to
stand accused of 'attacking the welfare state'. The
consensus allowed even contentious issues like the
introduction of comprehensive, non-selective secondary education and of earnings-related pensions to be
managed with considerable cross-party agreement,
despite some political posturing. The popularity of
these programmes underwrote this agreement.
    The British welfare state from 1948 onwards had
several fundamental characteristics not generally
found in other European countries. It was run directly by the state rather than through the financing of
private pension schemes, educational institutions and
hospitals; almost all staff in these fields were
direct public employees. It was comprehensive and
often universal in coverage; a flat-rate pension is
payable to all retired people, and free health care
and education are available to all. It is generally
free at the point of need; health charges (for drug
prescriptions and dental and optical treatment) are
a minimal source of funding; there is relatively
little private provision (over 90 per cent of health
and education is in the public sector, and the public
housing stock at 30 per cent of the total is high by
international standards). Most services are administered locally, but to national standards with little
discretion over policy.
    There are persistent areas of party conflict. but
there is not the same pattern of alternative ideologies as in economic policy. Conservatives have a
persistent preference for privatisation in health and
education, and especially housing, and for tight,
minimal cash benefits; but they participated in the

increase of social expenditure from 16 to 26 per
cent of GNP, with a doubling in real terms, between
1955 and 1980. They have introduced new benefits,
notably invalidity pensions in 1971, and even in
1983 they supported the index-linking of pensions
and the maintenance of health expenditure in real
terms. Labour has had a bolder vision of universal
benefits without stigma, but has more often used
these as a 'shock-absorber' of capitalism. with redistribution within rather than to the working-class.
Labour's policies have limited the growth in cash
benefits and frequently supported the interests of
welfare professionals and the middle class.

The more visible issues of party conflict have
included prescription charges, imposed in 1951 by
the Labour govenrment, withdrawn in 1965, but imposed again in 1968; and rent control on private
tenancies, relaxed by the Conservatives in 1957 to
the benefit of landlords and reimposed in 1965.
Particularly intractable was comprehensive education,
introduced by many local authorities in the 1950s
and 1960s, but encouraged generally in 1965 by the
Labour government through circular 10/65. This policy was reversed by the Conservatives in 1970, but
the proportion of secondary pupils in comprehensive
schools continued to rise to over 80 per cent in
1978, when the Labour government finally introduced
legislation to compel all authorities to reorganise;
this, too, was repealed in 1980.

Pensions has always been a major issue, because
the flat-rate pension established in 1948 was inadequate for workers not covered by occupational
pension schemes. In 1957 Labour abandoned its opposition to earnings-related pensions, and, after
abortive proposals by the Wilson and Heath governments, a comprehensive scheme was enacted in 1975,
to be phased in over 20 years. This will provide a
second, earnings-related pension for all workers not
contracted-out to an occupational scheme. The event-
ual settlement was generous to private pension
interests, reflecting the amount of negotiation and
compromise that had taken place. Meanwhile, general
pensions policy became more generous, as the elderly
took on greater political salience. From 1971 public
sector pensions were index-linked, and the flat-rate
pension has been uprated at least annually in line
with prices since 1973. The cost of the flat-rate
pension rose from 4 to 10 per cent of public expenditure between 1950 and 1980. Pensions are now one of
the largest single items of public expenditure.

Public housing support has also proved controversial, because of the balance of advantage between
public and private ownership. Subsidies of council
house rents are a visible symbol, attractive to
Labour and disliked by the Conservatives, and there
has been a difference in the level of support between governments; notably, the intention in the 1972
Housing Finance Act that local authorities should
move to self-financing 'fair rents' was overturned
by Labour in 1974. But public aid to owner-occupiers
through tax relief on mortgage interest is of comparable magnitude, and has not been challenged by
Labour. Politicians of all parties have shown an
increasing deference to the ideal of home ownership,
culminating in the Conservative policy in 1980 giving
the right to buy at a discount to public tenants.
This policy, a major vote-winner, contrasts with the
idea of housing as a public good still espoused by
Labour.

In recent years, both main parties have had to
come to terms with problems of the welfare state far
more fundamental than these issues of party controversy. Available output measures of social policy -
life expectancy, the number of low-income families,
the state of the housing stock, hospital waiting
lists, reading skills in schools - have failed to
show success commensurate with the resources applied
to the problems. Accusations abound that Beveridge
has been 'betrayed'; left-wing Labour thinkers have
been deeply critical of the policies of their governments; and Conservative free-choice theorists
have attacked the acquiescence of Conservative governments in a basically collectivist pattern of
provision.

Academic critics 'rediscovered' poverty in the
1950s and 1960s, pointing out that the numbers
claiming means-tested income support have increased
and that perhaps a quarter of households lack the
resources to participate in normal social activities.
New social problems have emerged like multiple dep-
rivation (the concentration of poor housing, unemployment and sickness), decaying public housing, and
over-prescription of drugs. New kinds of recipient
groups have become evident, such as one-parent
families, households with disabled members, immigrants and the homeless. Nor has public policy made
any great impact on the structure of inequality:
income and especially wealth are still skewed towards the upper groups, and what redistribution there
has been (mostly in wealth) has been from the very
rich to the merely rich.

The achievements of the welfare state are input-based, in terms of money, staff and coverage, but have not necessarily been translated into an output of real welfare. Much of the benefit of increased expenditure has gone into employment in health and education - up from 1.1 million to 2.9 million - where well-organised trade unions and professional associations have protected employee interests. Cash benefits, which have increased over four times in real terms, have not prevented more people slipping below officially-defined poverty lines: the number of recipients of means-tested social assistance (known as supplementary benefit since 1966) has trebled to over 3 million. Just as seriously, the starting point for income tax has slipped to under half of average wages, and national insurance contributions have increased: since 1975 they have been earnings-related rather than flat-rate, but are still a regressive form of taxation. The average household is receiving more and more social benefits, in cash and kind, but is paying more and more tax to finance them. The poorest families, especially where the head of household is in work, receive a poor deal from the tax-benefit equation, and much expenditure, especially on housing, transport and higher education, is of predominant benefit to the middle class.

Successive governments have been unable to get the operation of the welfare state right. In the 1960s and 1970s means-tested benefits proliferated - rent subsidies, student grants, legal aid and exemption from prescription charges - and two consequential problems were identified. The poverty trap resulted from the withdrawal of means-tested benefits as income rose but, in so doing, reached starting tax rate. The unemployment trap, or the 'why work?' syndrome, meant that the value of welfare benefits, especially for large families, could exceed the wages obtainable in employment. This reflected the relatively low level of child benefit (the flat-rate cash support paid to all mothers of infant and school-age children) but, more seriously, the low level of wages in the economy.

In these ways economic weakness fed back into social policy: welfare benefits are much easier to finance when there is real growth in the economy. Attempting to improve welfare services without regard to the economic circumstances that sustain them can only lead to recurrent, self-defeating crises. Moreover, welfare agencies that are near-monopolies and deal with clients in a weak bargaining

position (children, the sick, the elderly and unemployed) may become inefficient and even oppressive.
These are international problems which occur in systems based on both public and private provision; in
some countries they have stimulated a 'welfare backlash' in which electors call for lower taxes even at
the price of poorer social services. The British
welfare state has enjoyed remarkable public support,
but signs of crisis have begun to appear in the
1980s.

Until the 1970s, all parties shared a basic
faith in the purpose of the welfare state, given the
right organisation and level of resources. In these
respects, the 1960s were a decade of retooling under
both parties. There were major investigations into
education and the personal social services; new universities and polytechnics were established in order
to give a right of higher education to all qualified
applicants; the social assistance system was modified to make it more generous and less stigmatising.
In 1970 all the social worker services of local
authorities were integrated into unified departments,
and in 1974 the National Health Service went through
a similar integration under regional authorities. In
1974-5 the structure and boundaries of local authorities were reorganised and larger authorities created. Under the Heath government, the first attempts
were made to plan a system of negative income tax to
relate tax and benefits, though this was slowed by
the cost of implementation. Politically prominent
issues, like the Heath government's campaign against
welfare 'scroungers', or the attempt of some Labour
councils to resist the higher council rents imposed
by the 1972 Housing Finance Act, were in the long-
term much less significant than the undercurrent of
technocratic, consensual reform.

The reorganisations failed to win much public
approval and were widely associated with a decline
in services. At the same time - the mid-1970s -
there was the first real financial squeeze, especially by cutting capital expenditure on education
and later housing. Public sector wages were rest-
rained by the Callaghan government, and recruitment
of teachers was cut back as the number entering the
school-age population started to decline. Cash benefits, especially pensions, took an increasing share
of total social expenditure, but money to introduce
new types of benefit or assist those on supplementary benefit was tight. A new concern with results
and standards was evident, notably in Callaghan's

'great debate' on education. Although the 1974-79
government took some initiatives to help the disabled, and introduced child cash benefit which inc-
orporated child tax allowances, its policy tended to
be cautious and dominated by public expenditure
constraint. It reflected a growing scepticism about
the positive benefits of high investment in the
welfare state.

With the Thatcher government, an altogether
sharper policy divergence became evident. Education
expenditure was stabilised and housing cut sharply,
with a virtual end to mass public house-building.
Overall, however, although there were perceptible
cuts in certain fields, the trend of public expenditure was upward because of the recession (a larger
unemployment benefits bill) and because of demographic change (more pensioners). In 1980 the government abolished the earnings-related supplement
paid on unemployment and sickness benefit (without
any cut in contributions) and these benefits were
brought into taxation. Higher education was cut in
real terms, teacher numbers reduced, and rents for
public housing doubled. So far there have been
virtually no redundancies in the non-marketed public
welfare services (unlike the position in nationalised industries), but welfare professionals have
become beleagured groups, while the 'social' departments of central and local government have come
increasingly under the control of the Treasury.

Having borne the opprobium of cuts, the
Conservative government still had to cope with an
aggregate rise in real social expenditure. In a
welfare state based upon universal entitlement to
cash benefits and labour-intensive social services,
it is not possible to make substantial cuts without
restricting eligibility, holding cash benefit increases below the rate of inflation or declaring staff
redundant. The Conservatives have sought to avoid
this. On the health service they have been constr-
ained by manifesto promises in 1979 and 1983 to
maintain expenditure in real terms; in practice, the
increasing number of elderly people and the rising
sophistication of medical technology imply a weakening of provision relative to possible service
standards unless there is a real increase in expenditure, and so even here the policy has become associated with cuts. On pensions, the government is
committed to an annual increase in line with the
index of retail prices (having abandoned a link with
earnings enacted in 1974). Index-linking of pensions
has become a touchstone of faith to the 9 million

retired electors and seems to be sacrosanct, but
other benefits (like unemployment) are not covered
by the pledge and are vulnerable to real cuts.

In practice, cuts in unemployment or sickness
benefit or social insurance would have to be deep to
make any impact on the aggregate total. The dilemma
facing all governments is that the upwards pressure
from health and social security - as well as defence
- leaves little room for manoeuvre for any government seeking tax cuts as well as continuity with the
welfare state of the past. In some states it may be
posible to offload expenditure to subnational levels
of government; in Britain, where the centre in effect takes policy and financial responsibility even
for locally-administered services, it is not.

One Conservative policy response to this dilemma is privatisation, used in two senses - the
encouragement of private provision in housing,
health and education to relieve the burden on public
services, and the contracting-out of routine tasks
to private companies in the hope that they can be
done more cheaply. The policy of selling off council
houses to tenants at a discount has provided a capital windfall and reduced maintenance costs (though
the long-term benefit to the public sector is less
clear). Private hospital services and health insurance have boomed: in 1981, 7 per cent of the population were covered by insurance schemes. In some
areas, private insurance is becoming a necessary
part of acceptable health care for non-emergency
treatment, as well as an occupational perk. Health
authorities have been told that real growth in their
services will have to come out of 'efficiency
savings' from privatisation. But although it may be
preferable ideologically, contracting-out or increased charges to clients cannot yield the major
savings needed to resolve the government's cost dilemma.

The Labour Party response has lacked much of a
fresh approach to the problem, calling instead for
higher expenditure (including 3 per cent annual real
increases for health and the personal social services) and the discouragement of private provision.
In this, Labour expresses its close relation with
welfare professionals and teachers, and with public-
sector manual workers. The danger of this approach
is that, as with the Callaghan government, electoral
promises will prove too expensive and there will be
no clear notion of priorities; as costs increase,
the package is liable to fall to expenditure constraint.

Britain

The Liberals and the SDP have proved receptive
to revisionist arguments that the welfare state may
have created problems of its own, and that rising
unit costs and employee benefits need to be curbed.
They are attracted to comprehensive reform of the
tax/benefit interface through a negative income tax
system. The Alliance, and especially SDP leaders,
reflect the changing ideas of British elite groups
and the civil service. Once, the Beveridge ideal was
taken for granted and cherished with pride; now, its
achievements are doubted, and it is represented by
many as a drain on national productive potential and
individual economic responsibility.

New Issues

At the same time as the British polity began
demonstrably to fail to deliver the economic goods
and social welfare desired of it, it also began to
lose constitutional authority. The authority of the
state was questioned as it had not previously been
questioned in the present century. The new constitutional challenges took differing forms. The most
dramatic challenge arose in Northern Ireland, where
it was direct, violent in its methods and total in
its claim. Initially the challenge came from the
province's Republican (largely Catholic) minority,
but the British government's position was made the
more difficult because some Loyalists also began to
deny the legitimacy of British rule and began to
assemble their own paramilitary forces.

Northern Ireland had been ruled since 1922 as
part of Great Britain save that certain 'domestic'
issues were under the authority of a regional assembly. The assembly was dominated by a single political party, the Unionists, whose overriding aim was to
prevent their part of the United Kingdom becoming
detached from Britain and united with the rest of
Ireland. The Unionist Party was backed by the overwhelming majority of the province's Protestants, and
as Protestants outnumbered Catholics by a margin of
2 to 1 the Unionist hegemony was secure. In practice
the status quo was not entirely without support from
Catholics since the benefits to them as citizens of
the British welfare-state were considerably higher
than those they would have enjoyed as Irish citizens.
While the political representatives of the Catholic
population never accepted the status quo, they did
not use violence as a means to end it.

Inspired in part by the civil rights movements
of other countries and in some measure by the anger

of the increasingly large proportion of middle class
Roman Catholics at their political subservience, the
authority of the constitutional status quo was questioned at first by civil disobedience and then, in
the face of official and officially connived at violence, by violence. Quickly the initiative amongst
Catholics passed to the IRA and to even more extreme
groups.

British governments of both parties had had a
tacit agreement to keep Irish tensions out of British
politics. In practice this meant that as many issues
as possible were decided by the Northern Irish
(Stormont) government and its civil service in consultation with the Home Office. British politicians
feared that Irish problems could generate political
demands which cut across existing support for the
British parties, so that neither party could be
certain who would gain from the disruption. They also
suspected that these Irish problems were insoluable.
They were willing, therefore, to forsake the possibility of short-term gains and were more than happy
to handle these problems on a 'non-partisan' basis.
The troubles begun in 1967 did not disrupt this
agreement. Governments of both party tried to keep
the issue 'bipartisan', and in effect each has tried
the same policy repeatedly: to create a political
forum, an assembly of some kind, which would have
authority with both communities. While governments
have been reluctant to give such assemblies security
powers, the only substantial constraint has been the
unwillingness of any British government to abandon
its commitment to the majority community that the
province would remain within the United Kingdom. The
level of distrust between the communities was too
high - and it grew higher - for this policy to succ-
eed. For want of constitutional imagination, it
continues.

The other non-English parts of Britain, Wales
and Scotland, also posed a constitutional challenge
to the British state and this, too, made itself felt
first in the late 1960s. It was mostly peaceable,
however: the Welsh and, to a greater extent, the
Scottish people began to vote for their Nationalist
parties first at by-elections and then at general
elections. The Nationalist parties stood for the
independence of their nation from the British,
English-dominated, state. This position never won
the support of a majority of the Welsh or Scottish
people. Partly in response to this, the Natinalist
parties were prepared to support a half-way house to
independence: devolution. They sought the immediate

creation of directly elected parliaments in their
countries with considerable authority over domestic
legislation and over the executive. Ironically they
sought for themselves an arrangement very similar to
the one which had just broken down in Northern
Ireland.

Outwardly the British governments of the 1970s
responded to the nationalist challenges by offering
to meet them half-way. The Labour government of
1974-9 in particular actually put through legislation (subject to approval at referendums) for
devolved assembly. But the difference between the
response to Northern Ireland on the one hand, and
Scotland and Wales on the other, is not so great as
it first seems. In both cases British governments
began by acting in concert with the major opposition
party; the policy was 'bipartisan'. Also, in both
cases the policy was initially designed to concede
the minimum necessary to buy off support for the
Nationalist parties and return authority to the
British parties - the assemblies acted to preserve
the state and their own political position. But the
symmetry ends there. The interests of the two major
British parties no longer coincide entirely on
constitutional issues. Labour, with a high proportion of its vote in the north of England and in
Scotland, has an interest in keeping the Nationalist
parties at bay. The Conservatives have the luxury of
overwhelming support from the south of England,
where nationalism lacks any relevance. They can
therefore take a more relaxed attitude to the nationalist threat. Since 1979 they have renounced their
commitment to devolution.

In the case of Wales and Scotland the initial
policy was a considerable short-term success. The
widespread support which had been supposed to be the
main well-spring of nationalist voting evaporated by
the time the constitutional innovations were tested
by referendums. The Welsh rejected the proposal for
an Assembly by a margin of 4 to 1; the Scots approved their proposal so narrowly (and by a minority of
the electorate) as to make it impossible for the
government to proceed.

In the general election of 1979, which followed
the referendums by months, the Natinalists performed
poorly; in 1983 they lost still more support. The
British parties take some comfort from this return
to the status quo, but they do so at the expense of
having udnerstood and solved nothing. The well-
springs of nationalism are not discovered. It is
doubtful, too, whether the status quo could long

survive the return of any other than a Conservative
government. All the other parties are still committed to some form of devolution and all are deeply
indebted to the people who live in Wales and Scotland. They would be under pressure, which a Conservative government can avoid, to 'do something'.

One new issue which forced itself onto the stage
of British politics of the late sixties and after
was immigration and the integration into British society of the large numbers of African and Asian
immigrants who had arrived and settled since the
second world war. Public policy before immigration
became an issue was straightforward and simple.
Citizens of one Commonwealth country had a right to
travel to and work in any other. That hundreds of
thousands of Indians, Pakistanis and West Africans
would take up this offer had simply not occurred to
governments. To some considerable extent the traffic
in Black labour was encouraged by government and
other public agencies which needed an additional
workforce during the years of full employment. It
was only after there were substantial communities of
Black citizens in Britain that their presence became
an issue - or rather a series of issues.

The most contentious issue was the right of
immigration. Millions of Commonwealth citizens were
entitled to come to Britain: should there be no
controls? There was also an issue about the immigrants' right of access to the social services and a
related question about their right to equal access
to jobs in both the public and private sectors.
Finally, there were questions raised about the rights
of the communities of immigrants to preserve and
protect their own cultures.

Popular resentment against the incomers, stoked
by all manner of racist stories, rose after the boom
of the late fifties ended. From this point on, the
incomers were seen by many people as foreign competitors for scarce resources and competitors especially for jobs. The establishment was united in its
distaste for this sentiment but often fearful of
standing by its convictions in public, and it ret-
reated slowly in the face of popular pressure; in
general it was the Conservative Party which led the
retreat, Labour following in the van. The anti-
immigration speech of former Conservatrive minister
Enoch Powell in 1968 was a major watershed in this
process. The greatest concessions were made on the
right of immigration, which has been curtailed since
1962. On the other issues the establishment more or
less stood its ground and unrestricted access to the

social services remains the principle (and usually
the practice too). Legislation was passed against
discrimination in education and housing. There were
some important concessions: when Idi Amin threw out
Uganda's considerable Asian community in the early
1970s, the Heath government barely hesitated before
deciding to take them; on the other side, no amount
of official posturing by the central authority was
able to force some local authorities, like that in
Southwark (London) to give any publicly-owned houses
to Blacks (or even to allow other public authorities
to build in the borough).

The response to the new questions raised by
immigration and the popular reaction to it is inter-
estingly similar to the response to the new constitutional issues. Initially, the establishment tried
to keep the new issues out of party politics. The
effect, and to some extent, the intention of this
was to keep the issue out of popular control. So
long as party conflict was largely a two-sided
affair, with the leaders of both parties drawn from
the same elite, this was possible. The entry of the
new Nationalists and SDP/Liberal Alliance into the
fray complicates the competition but does not fundamentally change it. In the mid-1970s an openly
racist party, the National Front, did make considerable headway in some of the most depressed areas
where there are also considerable immigrant communities (especially Leicester and parts of North
London), but this support was not sustained and the
Front soon split into several factions. The continued ability of the large parties to keep race off
the list of issues which they dispute cannot be
guaranteed however: riots in the streets of many
large towns in the summer of 1981 could easily lead
to a reaction if repeated; and the introduction of
proportional representation (a primary goal of the
new SDP/Liberal Alliance) would open the way for
parties based on ethnic and anti-ethnic feelings.

The same mechanisms have been in force in regard
to the law on sexual behaviour and on private moral-
ity generally. Since the mid-sixties the law and
public norms on all these subjects have altered
fundamentally. Arguably, more British citizens have
been effected more personally by such changes in law
than by all the changes in economic and social policy. Capital punishment for murder is no longer
legal; abortion is; divorce is available on the
simple test of marital breakdown (and is the fate of
one new marriage in three); homosexual acts between
consenting adults is legal (even in Scotland) and

pornography is now freely available for sale. These
changes owe little to the main parties, the civil
service, prominent professional groups or, in some
cases, public opinion. They were all carried through
the House of Commons as Private Members' legislation
(in which neither government nor opposition take
part as such). Some of the reforms were unpopular at
the time of the change but have come to be accepted;
but some, such as the ending of capital punishment,
are not endorsed in opinion surveys.

In the case of law regulating the private lives
of citizens, the establishment, in periods when
parliament had Labour majorities, has acted in keeping with the prevailing international climate of
'informed' opinion. In other areas, it has not.
Consumerism, such a powerful force in the United
States, has made little impact in Britain. Governments stalled for years in face of strong evidence
on the harm caused by asbestos in industry; little
action is taken against smoking, and the rights
established elsewhere to protect consumers against
private enterprise or public bodies
finds no echo in Britain. British doctors still bury
their mistakes. Open government and the scrutiny of
public agencies, is poorly developed. In these matters, post-industrial issues as they are known on
the Continent, the British are backward.

Of course, these post-industrial concerns, like
consumerism and the peace and anti-nuclear power
movements, challenge the economic interests of entrenched parts of the establishment. The flowering of
the Campaign for Nuclear Disarmament in the late
1970s and 1980s was particularly uncomfortable to
government, as its message of peace had considerable
popular appeal and sought to puncture the mystique
of high politics in defence. The changes in the law
on private conduct challenged established and popular attitudes but few economic interests.

A final case - law and order - does not easily
fit into these categories. In general, the major
parties had assiduously kept policing out of party
politics. Both supported the general belief that the
British bobby was an admirable expression of British
values and one which foreigners ought to copy. The
consensus between the parties broke down in the wake
of riots of 1981 when the Labour spokesman on Home
Affairs took up the cause, previously the concern of
the far left, of police accountability. Policing and
criminal policy generally do not fit into the previous models because the situation has been changing
rapidly in many diverse ways. For one thing, the

prisons are badly over-crowded, so that any Home
Secretary has to urge a change in sentencing policy
which the aged English (not Scottish) judges resist.
On the other hand, the events of 1981 showed that in
some circumstances the police were not capable of
holding the line against rioters. In addition, the
occasional presence in Britain of IRA active service
units can not be dealt with in the normal way.
Finally, industrial conflicts in the seventies got
out of hand more than once and were actively stirred
up by some left-wing groups. All this serves to
question political authority and calls for a more
sophisticated response than a simple appeal to the
prerogatives of the state. So far, this response has
been lacking.

Conclusion

The most striking thing about British policy-
making and British government in the past two decades is its resemblance to previous periods. The
British not only have an old-fashioned constitution
with anachronisms like an hereditary chamber of
parliament, their system of government and their
policies have been becoming more bogged down as the
century grows older. The only major reform in this
period which has stuck is membership of the European
Community, and that is clung to more out of a sense
of the humiliation of withdrawal (and fear of economic reprisals) than of any sense of achievement.
Most of the other reforms to the machinery of government enacted in this period have been reversed or
seen to be failures. The operation and ethos of the
civil service remain unreformed; the system of local
government was outwardly reorganised, but to no
one's satisfaction; the health service was exposed
to managerial reforms which had to be simplified;
the trade unions are much as always - unique and
bizarre, quintessentially British in character; a
Central Policy Review Staff was tried, and dropped;
planning public finance by five-year constant-price
rolling programmes was attempted, and discarded;
devolution was offered, and spurned; an economic
planning ministry was created, and sat on; incomes
policies have been operated, and then abandoned;
comprehensive schools were created, and are now
being questioned; black Commonwealth citizens were
encouraged to emigrate to Britain, to find their
presence merely tolerated.
    The policies pursued by successive government in
economic, social and other fields have not been

notably successful, and in many crucial cases they
have been apparent and disastrous failures; and yet
the fundamental paradox is that far from leading to
an irrestible movement to change the machinery which
produces these policies, the opposite seems to have
happened. Support amongst the public for major constitutional and administrative change seems to have
evaporated. The failure over devolution was the most
embarrassing and most conclusive example of this.
It is fortunate for the European Community that the
referendum on membership was a decision about
whether or not to withdraw, not whether or not to
join, for the British public could then do what it
always does - opt for the status quo - and get a
change - membership. Pressure for change there
certainly has been, but it has been elite pressure
which has lacked the constancy to achieve a stable
result.
    Whereas many continental countries are moving
away from a class-based economic-issue politics towards something more like a post-industrial politics, Britain continues to fight the old battles.
This may be comfortable, it may attract wondering
foreign observers, but it cannot be a cause for
self-congratulation. It is, however, just arguable
that underneath all the movement in place there has
been a slow, fundamental rearrangement, expressed
by the exhaustion of one of the fundamental features
of post-war politics itself: the two-party system
seems in serious trouble. It is more than possible
that future elections will see the return of coalition governments and that these governments will
avoid the alternation of Conservatives and Labour in
power as the middle grouping, the Alliance of
Liberals and Social Democrats, shares in most governments. That would change things - or would it? For
the Alliance's credentials as an agent of change are
suspect. It is committed to a re-establishmjent of
the old, post-war consensus in social and economic
policy and aims to make only those constitutional
alterations (such as proportional representation)
which it hopes will entrench this consensus. An
appeal to a past that failed can not be a sound
basis for the future. That is why politicians of
both Left and Right are exploring increasingly radical solutions and inviting a cautious electorate to
join them on high-risk strategies which were incon-
ceivable in the recent past.

Chapter Three

FRANCE

J.R. Frears

"Capri, c'est fini!": That was the headline in
a Paris newspaper when the precarious state of the
franc and of the country's balance of payments in
1983 forced the government to impose severe restrictions on currency for foreign travel. In a way it is
an epitaph for a whole period in which France was a
paradise island of dynamic growth, modernisation,
dramatic improvements in living standards and welfare, harmonious consensus over political institutions, and a marked absence of the social tensions
reflected in terrorism, racism, and violence that
characterised many European countries in the 1970s.
Critics of the present French government have no
difficulty in putting a date on the turning of the
tide: 10th May 1981, when the socialist leader
François Mitterrand won the presidency. While it is
true that the early measures of his new government
did have an adverse effect on the competitiveness of
industry, in a deeper sense paradise had already
been lost long before. The slowing of economic
growth, the rise in unemployment, the growing strain
on social security budgets, the realization that
France was not going to achieve the breakthrough to
a Japanese-type of leading high-technology economy -
these did not start on the 10th May 1981. The 1970s
left some terrible problems for Western economies -
in particular the multiplication by fourteen of the
price of oil and the shift in the manufacture of a
wide range of quite sophisticated industrial products
from advanced countries to less developed countries.
How, then, economically, socially, politically, is
France facing up to the difficulties of the 1980s?

France

Economy

As one author well put it: *"Economic growth is nice. It lubricates tensions and dissolves conflicts. It is relatively easy to manage and delightful to score."* One could add that it gives self-confidence and assurance to governments. Military spending, welfare spending, prestige spending, infrastructure spending - all can be afforded, all demands met from a growing budget. An advanced and expanding economy is the basis of high living standards at home and diplomatic success abroad. But, "untune that string, and hark what discord follows". The checking of economic growth has put tremendous strains on the budget (especially in social security), produced two million unemployed, caused discontent to surface in the form of protests against immigration or action by interest groups to defend their *situations acquises*, forced the government to borrow internationally to save the franc. It has led to a form of political debate terribly familiar to the British but unknown in France for a quarter of a century: arguments about rates of tax in the budget, arguments about percentage pay increases for different groups and obsessive attention to industrial disputes, arguments about cuts, talk of private insurance replacing some public health provision, a propensity to blame everything on the government of the day (the British, curiously enough, have suddenly stopped doing this). Anxiety and doubt have replaced the rather arrogant self-confidence of the 1960s and 1970s. Diplomatic initiatives have lost the spectacular, if empty, quality they used to possess. Dashing military interventions like the action of President Giscard d'Estaing in Zaïre in 1978 have become somehow unthinkable. Chad in 1983 was a very different business.

The omnipotence of executive power, concentrated in the hands of the President, has been the most striking feature of the Fifth Republic system of government. Institutions like parliament and the courts, other checks and balances like the mass media and interest groups, have all be ineffective to restrain the gigantic power of the French state. The opposition had never won an election since 1958. The 1980s have seen the end of the omnipotent executive. The 1981 elections showed that even a President, with all his concentration of power and patronage, could be removed by the people. That has administered a great and salutary shock to the system and reduced executive power to a more human scale. But it is really the end of rapid economic growth that has undermined omnipotence, because executive power now faces much more discontent and pressure over the distribution of scarce resources - always a severe problem in a democracy. From an examination of the economic problems of France in the 1980s, one may move on to see how they have affected the system of government itself, in particular the stability of political institutions as less easy times come.

France has, as everyone knows, been one of the economic miracles of the post-war period. Between 1949 and 1973, apogee of the boom, annual production of goods and services increased three and a half times. After the oil crisis of 1973, growth began to slow down but the record is honourable, superior on the whole to other developed countries in Europe and in the world generally. In the period of rapid growth, France reached the rank of fifth industrial power in the world and fourth exporter until overtaken by Great Britain in 1980.

Those were the days of the omnipotent state. The French state has a long *dirigiste* tradition and there never was in France a classic period of industrial growth based on capitalist enterprise. Consequently, in the post-war period, particularly after political stability was secured in the 1960s, it was well placed to direct economic growth according to national objectives defined by the planning machinery, to provide the capital, the regulatory power, and the expertise in the form of its technocratic public service elite, and to arbitrate any conflicts between employers and trade unions which were weak, divided and, in any case, rendered docile by the fruits of economic growth. French industry, traditionally protectionist, was made to face up first to the opportunities and the threats of membership of the European Community and then to the challenge of world trade generally when the internationalization of economic life reached maturity in the 1970s. The policy pursued was one of 'national champions' - the goal of one (or at the most two) internationally competitive French firms in every important industrial sector, especially the most advanced ones. This has been achieved most spectacularly in the automobile industry, in the field of military equipment, and in the nuclear power cycle from enrichment to treatment of waste. From the mid-1970s, however, things have started to go wrong. Two are world problems - the energy crisis and the arrival of the 'new industrial countries'. A third

is the ending for France of what has been called
'the advantages of backwardness'. A fourth is the
huge balance of payments deficit that has developed.
A fifth is the structural deficit in public finances
as rapid economic growth ends and social expenditure
accelerates. A sixth is the effect on industrial
competitiveness of government policy since the
election of François Mitterrand.

Energy

The effect of the immense increase in the price
of oil on the economies of Western countries is too
well known to need repetition. OPEC quadrupled the
price of oil in 1973 and trebled it again between
1978 and 1981. It was $2.5 a barrel in October 1973,
$36.9 in January 1981. This hit France particularly
hard for three reasons. Firstly, France was exceptionally dependent on oil (virtually all imported)
for total energy needs (66 per cent in 1973). France
has been more effective than most countries at doing
something about that, but even so the burden is
considerable, a bill of Frs 125bn in 1980. Secondly,
the bill had to be paid in dollars which had risen
very sharply against the franc in the early 1980s
from around 5 Frs to 8 Frs. Thirdly, whereas the
German workers and the Japanese were prepared to
accept the oil price increase as a once-off cut in
living standards, the French, like the British, were
not willing. France therefore, like Great Britain,
felt the full force of inflation in the 1970s and
lost out to their principal industrial competitors.
This is interesting because many writers have
commented on the ineffective and fragmented character
of interest groups in France and their incapacity
(compared with interest groups in the Anglo-Saxon
democracies) to impose a veto on public policy.
However, as events since 1981 have confirmed,
interest groups in France are not so ineffective
when it comes to defending *situations acquises* .

The French response to the energy crisis has
been a characteristically technocratic operation to
which the French state is very well adapted. Firstly,
the state has controlled since the 1920s the import-
ation, refining, sale, and prices of petroleum
products, and the two principal French oil companies
(CFP and ELF/ERAP) are controlled by the state.
Secondly, an enormous nuclear power programme was
launched - by far the largest in the western world.
The aim was to develop at five thousand megawatts a
year until, by 1985, half of the nation's electricity

was to be produced by nuclear power (one-fifth of total energy requirements). Thirdly, the state embarked on a vast programme of arms sales to oil producing states in order to guarantee supplies and pay for them.

The nuclear power programme is a very good example of the technocratic style of political decision-making. The decision to 'go nuclear', to do so on the basis of the *francisation* of the apparently exportable Westinghouse PWR system, the creation of a monopoly based on EDF (Electricité de France), sole purchaser, and Framatome (the Westinghouse licensee - in which CEA, the Commissariat à l'Energie Atomique, has a large holding), the creation by CEA of COGEMA to be responsible for the production of combustible material and its disposal in retreatment plants, the development of uranium mining in France and in French client states in Africa, the development of the Super-Phenix fast breeder reactor at Creys Malville due to go into production in 1983-84 - all this has been a great technical success. However, as the Commission du Bilan (set up in 1981 to assess the previous government's record) remarked, *"the nuclear programme has only been embarked upon and achieved at the price of a reinforcement of technocratic trends"*. Everything was decided by government regulations, not legislation discussed in parliament. The system of development control in France, furthermore, does not give the public a chance to form pressure groups that will get a hearing at a judicial-type public enquiry. The British middle class, adept at organising objections to developments it does not like, would be astonished how easy it is for the French state to order the construction of airports, coal mines, highways or nuclear reactors. Consequently, the conception to completion time for a French nuclear power station is about six years, compared to twelve anywhere else; and the big constraint on development that has made itself felt in other democracies like USA, West Germany, Sweden or Great Britain - namely public objections to and anxiety about nuclear power - has simply not materialised. There have been some environmental protests, but nothing on the scale of other Western countries. The only one that threatened to get out of hand arose from the Breton coast village of Plogoff. Fortunately, a promise to cancel the Plogoff power station was included by Mitterrand in his election campaign, so that was defused. Indeed, environmentalists and socialist objections to an 'all-nuclear' policy have played a part in the

slight slowing down of the nuclear programme since
1981.
   The French nuclear programme could lead to some
big difficulties. There have been cracks in reactor
cores, giving rise to some anxieties about safety;
there is the danger of proliferation (Israel in 1981
had to bomb a reactor sold by France to Iraq); the
world-wide export market for nuclear power stations
(a major consideration of French government policy
in opting for the PWR system) has completely dried
up and forecasts of the number of reactors to be in
operation world wide by the year 2000 have fallen by
a factor of 5. Finally there is considerable anxiety
about the escalation of costs. Some believe that the
nuclear programme could burden France with the highest energy costs in the world. The costs of operating the fast breeders and reprocessing plants appear
to be rising alarmingly, and the Commission du Bilan
expressed disquiet about power station construction
costs. The EDF was criticised by the Commission on
Competition for paying too much for monopoly-
supplied nuclear power stations.

The Newly-Industrialised Countries

   The second big world problem which has created
difficulties for the French economy is the arrival
on the scene of the newly industrialised countries
- Brazil, Singapore, Hong Kong, South Korea, Spain,
etc. This is the familiar problem of the 'product
cycle'. Products originally manufactured in the most
advanced countries shift towards less developed
countries. It starts with textiles and shoes ; then
it becomes radios, watches and cameras; then in the
1970s steel, ship building, automobiles and tele-
visions began more and more to be manufactured in
the newly industrialised countries. The really
successful advanced economies are the ones that have
anticipated the trend by getting into higher and
higher technology - USA, Japan and West Germany.
Protectionism is not enough - it may be possible to
mitigate the effects of imports on one's own domestic market, but it is the falling share of
exports as one's customers' markets are penetrated
that does the damage. France has been able to limit
Japanese car imports to 3 per cent, but the French
share of the automobile market in other countries
has been hit by Japanese imports. The French problem
is that it has not quite achieved a rapid enough
conversion out of the declining traditional industries into the sunrise industries (electronics,

information technology, robotics, and so on). Big
efforts are being made - the Socialist government
has doubled the budget for research and technology -
but France entered the 1980s, in the words of the
civil service Director General for Industry, as
"strongest of the weak or the weakest of the strong" -
suspended between the USA-West Germany-Japan group
and the Great Britain-Italy group.

The 'Loss of Backwardness'

The next problem for France in the 1980s that
requires examination is the loss of the 'advantages
of backwardness'. The original goal of economic
planning in France was modernisation, especially the
rapid growth of the industrial core of the economy.
It was a brilliant catching-up operation. It was not
just state *dirigisme*, although the public service
elite played a leading part. The optimistic climate
of growth produced some managers and business
leaders of exceptional quality in the private sector
- many of them, of course, trained in the public
service training schools of Polytechnique and ENA
(National School of Administration) and drawn from
the highest civil service corps.
    What was the state to do now? The late 1970s saw
a relaxation of its role in some areas of economic
life, e.g. prices, and increased reliance on market
forces to restore profitability, an official ideology of no longer insulating uncompetitive 'lame
ducks' (though more honoured in the breach than the
observance). The 'Interim Plan', however, introduced
by the Socialist government soon after arriving in
office, having scrapped the previous administration's
8th Plan, criticised the *"disengagement of the state...
a policy that was in fact liberal because it preferred adaptation through the market to the deliberately willed action of
the state"*. In fact, as the Interim Plan tacitly
admits, the *dirigiste* approach of restructuring the
economy away from declining sectors and into advanced
technology has continued, because   the big,
advanced-technology successes had been *"the result of
public enterprise or public policy"* - the high speed train
(TGV), the Airbus, telecommunications equipment,
nuclear technology. Strangely enough, military aircraft and equipment, one of the greatest French
successes of all, is omitted from the list. Nevertheless the list, now completed, shows how France has
tried to specialise in that vast and terrifying field
of the modern world economy - state-to-state dealings,
or what the Americans call megabuck contracts:

equipping an airforce, supplying whole telecommunications or transportation systems for a developing
country, nuclear reactors or turnkey contracts for
industrial plants. Defence has been the most successful field - especially sales to the oil producing
states in the Middle East 'crescent of crisis', the
dangers inherent in which need no emphasis. In 1978
the defence correspondant of *Le Monde* complained
that the state was doing too much, arranging design,
promoting sales, manufacturing, financing, insuring,
controlling companies, providing after-sales
service, training users, and getting into shady
deals with dictatorships of all sorts. France is the
leading exporter of arms after the USA and Soviet
Union: some 300,000 people (4.5 per cent of the
workforce) are employed in arms production and arms
exports in 1980 were Frs 23.5bn.

The Socialists have reinforced *dirigisme* by a
considerable extension of the public sector. The
1981/82 programme of nationalisation involves steel
(already effectively nationalised by the Barre
government in 1978), that part of the banking sector
that was not already under the state control, except
small and foreign-owned banks, five of the largest
industrial corporations (CGE - electrical engineering, ship building, telecommunications; Pechiney-
Ugine-Kuhlmann - chemicals, aluminium; St Gobain -
glass, engineering, electronics; Thomson-Brandt -
electrical; Rhône-Poulenc - chemicals, pharmaceut-
icals), the leading firms in the armaments sector
(Dassault and MATRA), and the French branches of
certain foreign multi-national corporations (CII-
Honewell Bull in computers; Roussel-Uclaf, a subsidiary of Hoechst Chemicals; ITT-France in telecommunications). *Dirigisme?* Nationalisation was to
be on the Renault model - the firms were to be run
on commercial lines with little state intervention.
The government's letter to the newly appointed
chairmen began, *"You will strive, first of all, for
economic efficiency..."*. However, if one looks at the
preamble to the nationalisation law or the references
to it in the Interim Plan, one finds rather more
direction. The Council of Ministers' statement in
September 1981, announcing the plan for nationalisation, declared that from 1974 to 1980 investment
in public enterprise had increased by 91 per cent
while in the private sector it had declined by 5 per
cent. *"The extension of the public sector ...will
permit the catching up and the modernisation of our
whole productive structure ...The nationalised banks
will have the task of supplying finance better*

adapted for these modernisation needs". They were
set many social objectives as well - from providing
more jobs (increased output, shorter working hours,
recruitment of young workers, etc.) to concern for
protection of the environment and being innovators
in giving more responsibilities to workers and
decentralised management.

It is too early to say whether this attempt to
give a new impetus to modernisation through an
enlarged public sector will work. The first results
are not encouraging. The eleven nationalised firms
in the competitive sector, which had losses of Frs
17bn in the 18 months up to mid-1982, lost a further
Frs 5bn in the next six months. The plans they sub-
mitted to the government for 1982 requested Frs 50bn.
They faced two sorts of difficulty. One is the threat
to competitiveness from the ambitious social programme described below. The other is the obstinate
lack of commercial success in sectors judged to be
vital. We have already noted the collapse of the
world export market for nuclear power stations.
Another example is the extreme difficulty the French
have had (for many years before nationalisation) in
establishing a foothold in big computer systems. In
the 1960s, partly for prestige, partly from the
perceived insecurity of too great dependence on
America, they tried to develop, through the Plan
Calcul, an independent 'national champion' based on
French technology. It failed. Under President
Giscard d'Estaing this was dropped in favour of the
francisation of an American product line (as with
Westinghouse in nuclear power). CII-Honeywell Bull
was created, which proved more viable, relied mainly
on Honeywell computers from America, but in 1981
recorded a Frs 430m loss and vast financing needs
for the future. The information technology industry
in France thus finds itself back at square one. It
is hard to see that new talk of national independence
and the loosening (though not severing) of links with
Honeywell through the nationalisation of the
American company's holding will really help. Inevitably, the state, faced with the dilemma of either
closing down in big computers or giving an enormous
open-ended injection of capital, seems to have
adopted the half-way solution of 'paying the monthly
bills'.

Overseas Trade

Another major problem for the French economy in
the 1980s is the catastrophic deterioration of the

country's balance of payments. In all but two years since the oil explosion in 1973 French overseas trade has been in deficit, reaching the alarming figure in 1980 of over Frs 60bn. The Socialist government's dash for growth, began when it came to office in 1981, made matters worse because higher take-home pay and benefits led to higher imports. The 1982 balance of payments deficit was Frs 93bn (some 2.5 per cent of GNP). Jacques Delors, the Finance Minister, was forced to seek large international loans in 1982 from the Saudis and from the Euromarket. In 1983 the European Community made a further loan, but the price was an austerity package of Frs 65bn of cuts - which provoked the *"Capri, c'est fini"* remark with which this chapter began.

Analysing French overseas trade, the Commission du Bilan in 1981 found that, excluding military sales, the deficit with the three leading industrial countries (USA, Japan, West Germany) had grown to Frs 48bn by 1980 and with OPEC to Frs 65bn. France was in deficit to OECD countries as a whole. The French deficit with EEC countries reached Frs 64bn by 1982. Agricultural exports had been extremely successful, so the basic problem was industrial. France still has a surplus of exports over imports in manufactured goods, but in 1982 it was down from Frs 43bn to Frs 19bn - with which to begin to meet an energy deficit in 1982 of Frs 138bn. The industrial trade problem is illuminated by the products which achieved the best surpluses in trade with the developed countries of the OECD: the top six are cars (Frs +16.1bn), wheat (+4.6), wines -*appelation controlée* (+3.4), spirits and liqueur (+2.6), live cattle (+2.6), other cereals (+2.2).

Budget Deficits

The next problem associated with the end of rapid growth is laconically summarised in the Commission du Bilan report: *"The rupture of the high -growth economy has created in France, as in the other western countries, the conditions for a structural deficit in public finances."* Resources stop growing but public expenditure is not easy to restrain: indeed declining growth accelerates public expenditure because the demand for unemployment benefit and social security increases so dramatically. France, like most countries, has had a budget deficit - a public sector borrowing requirement that the British are so used to arguing about - since the oil crisis hit growth. It averaged around Frs 30bn

(not counting social security) for the second half
of the 1970s and has risen in the 1980s. For the
1984 budget President Mitterrand undertook to keep
the public deficit (including social security) down
to 3 per cent of GNP (around Frs 125bn).
   The French state has an extra difficulty in
balancing the books because of the unusual way
social security is financed. Sickness benefits and
the cost of medical and hospital treatment, retirement pensions, family allowances, and unemployment
benefit are all (or virtually all) financed by
employer and employee contributions which do not
form part of the state budget. This means that
although 'compulsory levies' (direct and indirect
taxes plus social security contributions) are high
in France (they have grown to nearly 45 per cent of
GNP for 1984), a far lower proportion comes from
direct taxes than in any other OECD country.
   The French system of taxation is therefore much
less progressive than in other countries. Since most
benefits have a ceiling, contributions under some
headings have a ceiling also. This means that a high
earner pays a much lower percentage of income in
contributions than a low earner. The system also
poses a big problem for firms, since contributions
split almost 3 to 1 between employer and employee.
This means that it is extremely costly to take on a
worker - which may have favoured capital investment
and productivity in the period of rapid growth.
Contributions have continually to be raised in order
to meet the rapid increases in the cost of health
treatment, the rises in benefits decreed by government in line with inflation, the increased demand
for benefits through rising unemployment and so on.
Consequently, firms find their costs rising and
their reluctance to cooperate in schemes to employ
more workers growing.
   Such has been the resistance to higher contributions by employers and employees in recent years
that an alarming and rapidly growing deficit in the
health and social security system has developed.
Every Minister of Health and Social Security in
recent years has declared that his primary objective
is the elimination of the deficit. The exception was
the first minister of the Socialist government,
Nicole Questiaux, who declared she was not going to
be Minister of Accounts, but her successor has made
a particular effort to balance the books. All kinds
of measures are being tried - special taxes on
tobacco and alcohol, special 'solidarity contributions' by employees of the state who have safe jobs

and are exempt from contributions, 'hotel' charges
for hospital patients, increased employer and
employee contributions to the unemployment insurance
scheme, extra taxes, including a proposal in July
1983 for a special 2 per cent levy on all incomes.
One of the most radical ideas is the transfer over
five years of family allowances to general taxation
under the state budget. There have been angry
scenes. Almost every proposal has been condemned by
unions and employers.

The Mitterrand Measures

Finally, under the heading of economic problems,
the only one of the six discussed that can be blamed
entirely on the Socialist government: the effect on
competitiveness of the social policies undertaken
since 1981. Industrial costs have been hit three
ways. The first is through increased employer contributions to social security, required largely to
meet increased benefits. One of the early acts of
the new government was a 20 per cent increase in the
minimum pension, 25 per cent increase in family
allowance and 50 per cent in housing allowance, and
the ending of the system under which patients paid a
proportion of medical costs. The second factor was
the effect on taxation and interest rates on public
expenditure generally: direct creation over two
years of 210,000 public sector jobs in fields like
health and education (promised by Mitterrand in his
election campaign); a 27.6 per cent increase in
public spending for the 1982 budget over 1918 (itself 16.4 per cent up on 1980); a budget deficit for
1982 of a record Frs 95bn (2.6 per cent of GNP). The
third factor was a whole range of social policies
designed to improve life at work and mitigate unemployment by sharing out work. Retirement at 60 and
a fifth week of paid holidays were introduced, and a
reduction of the working week, towards the objective
of 35 hours, was inaugurated by a cut in the legal
maximum (excluding overtime) from 40 to 39 hours.
The legal minimum wage has been steadily increased -
and by more than under the previous government.
     The greatest set-back involved the reduction of
the working week. The original hope was that workers
would not get 40 hours pay for 39 hours work but that
the time 'freed' would be used to take on more
workers, while giving those already in employment
more leisure. A series of strikes soon made it clear
that the 'acquired right' of 40 hours pay was not
going to be given up, and President Mitterrand

reassured workers there would be no loss of pay.
Consequently, the measure had a direct incidence on
industry's labour costs and did very little to
promote a reduction in unemployment. So concerned has
industry felt about the increased costs it has had
to bear that a committee was set up to quantify them,
reporting in July 1983. The two sides, government
and industry, could not agree, so separate calcula-
tions were listed. The government considered that
its policy would increase industry's costs in 1983
by Frs 28bn; industry considered the correct figure
to be Frs 62bn (with Frs 31bn attributable to the
reduction of working hours alone). Either figure is
quite large enough to have a serious effect on
profitability and hence investment.

The attempt at a Keynesian reflation of the
economy through public expenditure as a stimulus to
demand has helped to produce a higher rate of
inflation in France than in most of its competitor
countries. In 1982 inflation in France at 11.8 per
cent was still above the 1973-80 average of 11.1 per
cent, while in the EEC and OECD countries as a whole
the rate of inflation had fallen perceptibly. This
is certainly not going to help the French nation
overcome economic problems such as the balance of
payments and industrial exports - indeed the attempted dash for growth in 1981-82 drew in more imports.

The nationalisation measures may also be having
their effect on the competitiveness of French
industry. They did have social as well as economic
objectives. *Le Monde* published an interview with
Jean Gandois who was head of Rhône-Poulenc when it
was private, and maintained in that post when it was
nationalised despite union objections, only to resign
a few months later. There were, he said, two concep-
tions of the nationalised firm. The first, which he
understood to be that of the Industry Minister of the
time, Pierre Dreyfus, former head of Renault, he
agreed with. It was that *"the public enterprise is
distinguished from the private by three differences:
it must take more long-term risks; it must practice
the same social policy as the private firm while
recognising a duty for social innovation; it must
respond to the requirements of the state, provided
these do not penalise it commercially or divert it
from its main tasks"*. The second conception, which
he evidently felt was beginning to prevail, was that
*"the mission of the nationalised firm is to achieve
the economic and social objectives of the government.
That can lead to investment without being sure that
a market exists, thereby creating the temporary*

*illusion of new jobs; to the sacrifice of financial
balance for over-ambitious investments; to major
concessions over productivity in order to favour the
maintenance of employment; and to risky experiments
with the distribution of power in the firm".* He
concluded: *"little by little, the firm from the
competitive sector is transformed into a sort of
public service which the state must subsidise to
enable it to survive".*

Society

Insecurity

The social problem that most obviously derives
from the economic difficulties discussed above is
unemployment. In 1983 this stood at around the 2
million mark (9.4 per cent of the workforce), having
previously increased at about the same pace as most
developed countries (much less than in Great
Britain). Over a quarter of those out of work have
been out of work for more than a year. This has
undoubtedly contributed, as in other countries, to
the rise in crime, especially by young people, to a
climate of insecurity, and to an increase in racial-
ist incidents. It has not, in France, any more than
in the traditionally stable Great Britain, led to
the kind of unrest or violence that might threaten
political institutions.

*"How much of the rise in minor street crime must
be attributed"* asks the Commission du Bilan in its
section on civil liberties *"to poverty and frustration arising from youth unemployment? One factor
is certain, the multiplication of incidents which
affect daily life - burglaries, muggings - has
progressively created a psychosis in wide sectors of
public opinion: fear of the inhabitants of large
estates on the fringes of out cities, fear of the
corridors in the metro at off-peak hours, all that -
much more than serious professional crime, which has
remained static, much more than the first manifest-
ations in France, from 1977 on, of an international
terrorism which is far worse in neighbouring
countries - is at the origin of a feeling of insecurity that the mass-media and the statements of certain
political figures...have amplified."*

It was this climate of insecurity that caused
President Giscard d'Estaing to refrain from asking
parliament to abolish the death penalty, that
produced the 'legitimate defence' association,
founded to support those who took the law into their

France

own hands against intruders or burglars (their
actions resulting in a reported 21 deaths) and the
Securité et Liberté law of 1980. This law, introduced by the Minister of Justice, Alain Peyrefitte,
'reinstated the certainty of punishment' (in other
words removed from the judge the power, whatever
the individual circumstances, to give lighter or
suspended sentences for certain serious crimes). It
also speeded up procedure by giving the state's
Chief Prosecutor (the *Procureur de la Republique)*
the power to cut out one of the judicially-
controlled investigation stages by which serious
criminal cases proceed to the assize courts;
retained most of the much criticised procedure for
direct trial of an accused 'caught in the act';
lengthened the period in which the suspect could be
held by the police without access to a laywer; and
authorise police identity checks. In short, more
securité than *liberté*.

The Socialist government elected in 1981 was
committed to a number of judicial reforms of a very
different direction to this prevailing wind. The
death penalty was abolished. The State Security
Court, a relic of de Gaulle's counter-terrorist
measures at the height of the Algerian crisis in
1961-62, was abolished. The law whereby anyone
attending a demonstration or protest is considered
responsible for damage done during it, was abolished.
The *Securité et Liberté* law was eventually abolished
too, but events at first were not entirely prop-
itious. There were some serious terrorist outrages
in Paris in the summer of 1982, notably the bombing
of a well known Jewist restaurant in the rue des
Rosiers, and this rising crime wave was blamed on
the liberalism of the government. A minister was put
in charge of the fight against terrorism and identity checks were retained.

The government, in this whole field, is navigating in waters of hostile public opinion which the
opposition parties are doing their utmost to
exploit. In June 1983 there was a very serious
episode - a demonstration by police against the
government after the deaths of two more police
officers in Paris. Trust between police and citizens
has never been very strong in France, and it seems
to be getting worse. There are said to be tough
suburbs into which the police cannot go at all without being attacked. The problem should not be
exaggerated, however. France has, since the end of
the Algerian war, had far less terrorism than other
European countries - nothing domestic on the scale

of the IRA, the Italian Red Brigade or the Baader-
Meinhof gang in Germany. The worst is what is taking
place today in Corsica. Paris has suffered from some
international terrorism - Armenian, Palestinian,
Iranian - because Paris is an international centre,
but it has been no worse than anywhere else. More
worrying is the persistent series of anti-Jewish
outrages - possibly Palestinian in origin, possibly
straight anti-Semitic - which have continued since
the 1980 bomb attack on the synagogue at the rue
Copernic.
    Economic recession is always a bad time for
immigrants. They occupy the most vulnerable unskilled jobs. They are condemned as scroungers when they
become unemployed; their housing and education are
usually the worst and suffer from cuts when cuts
come; they are blamed for crime and for 'taking our
jobs'. The recession has seen an increase in racist
incidents - though the French record for racial
tolerance is arguably   more honourable than that
of the Anglo-Saxon countries. The principal ethnic
minority is North African - 750,000 Algerian and
250,000 Moroccan immigrants, plus their 300,000
children born in France and therefore of French
nationality. Like crime, racial prejudice at times
of insecurity is a tempting theme for right-wing
opposition parties. Jacques Chirac, for instance, in
July 1983 evoked the spectre of Pakistanis being
expelled from Great Britain and sent to France - a
most unlikely scenario.
    The principal government response to the problem
during the Giscard Presidency was to stop immigration of workers from non-EEC countries and of families of immigrants already in France. After protests
from the Algerian government, whose citizens were
the most affected, and a ruling by the Conseil
d'Etat, the restrictions on families were lifted.
There was a vigorous, but largely unsuccessful,
programme of repatriation grants for immigrants. Had
it not been for Algerian government protests and a
serious threat to Franco-Algerian relations, a policy
of systematic non-renewal of expiring five and ten-
year residence permits would have resulted in the
expulsion of several hundred thousand Algerians. Had
it not been for the Constitutional Council, the 1980
law would have introduced, in addition to measures
for expulsion without appeal, 'administrative intern-
ment   pending departure for any immigrant who had
no work permit or whose presence 'represented a
threat to public order'. The Socialist government
has abrogated the 1980 law, and made it possible for

66

the large number of illegal immigrants, already in the country for a long time and very prone to exploitation by their employers, to regularise their situation. Illegal immigration remains a problem, however, to which public opinion and the government is very sensitive.

Education

The debate on education reform is another theme of French social life that is linked to the economic crisis. The Minister of Education, Alain Savary, introducing a law to reform higher education in parliament, said he did not consider that French universities *"were at present in a state to make their contribution...to our country's battle for its economic, social, and cultural development...To stay in the leading group of developed nations we have no alternative but investment in our human resources"*. He quoted President Mitterrand: *"My determination is to arm France with a human capacity equal or superior to any in the world"*.

The emphasis in the last decade has been on 'elitist egalitarianism': to raise the cultural and educational level of the population by improved opportunities but to retain incentives and special advantages for the brightest and most motivated. As a result of the Haby reforms in the mid-1970s (Ministers of Education like to give their name to a law of structural reorganisation), secondary school children all now go to a mixed-ability *Collège Unique*. The first two years there is called an 'observation cycle', the next two an 'orientation cycle'. At this stage there is supposed to be intensive guidance for children to orient them in the most suitable direction for their particular talents. This system has been greatly criticised. It was supposed to ensure that all children receive a good secondary education. In fact, the schools shunt the non-academic into dead ends called 'pre-professional classes' or 'preparatory classes for apprenticeship'. These classes were not intended to be dead ends, indeed they were specifically meant to have a technological bent, but the Ministry (supported by the teacher unions) recruited their teachers very rapidly, often by promotion from primary schools and without science training or even, in many cases, higher education. Too many working-class children, moreover, are prematurely oriented towards the *Lycées d'Enseignement Professionel*, which provide vocational training (e.g. in catering) without

67

giving the broad secondary education envisaged by
the Haby law. It is now generally felt that *"the
Collège Unique does not exist"*. Like the comprehensive schools in Great Britain, they vary enormously
according to factors like the neighbourhood in which
they are located.

To many British readers this all will sound
unduly self-critical. Education in France has some
very positive features: from the excellent quality
and almost universal provision of nursery education,
to the high percentage of the 16+ age group in full-
time education. Over 50 per cent of 17-year olds
and 15 per cent of 20-year olds were receiving full-
time education in 1979, about the same as West
Germany but much higher than Great Britain. For the
brighter children who are 'oriented' to schools that
prepare for higher education, the baccalaureat
exams, taken at 17 - 18, involve a wider range of
subjects than is taken by A-level candidates in
Britain. It has become increasingly fashionable for
children of the professional middle-classes to take
a baccalaureat with an orientation to maths and
science, though without neglecting the humanities,
rather than the traditional orientation to philosophy and literature, because that is the way into the
engineering and business schools (parallel to the
universities) which open the way to the top jobs in
the public services and industry.

The new Savary law on higher education (another
appélation ministérielle) addresses itself to inequality of access in the same way that the Haby law
hoped to do in secondary education. The problem in
France is not, as in Great Britain, restrictions on
the numbers allowed into higher education (everyone
who passes the baccalaureat is entitled to enrol at
a university) but that the vast numbers of first-
year students receive virtually no teaching. The
drop-out rate is around 60 per cent. That way the
universities arrive at manageable teaching numbers
for the second and third year. One of the main
objectives of the Savary law is to give a wide
pluridisciplinary range of courses, including
science and languages (which could mean computer
language) to this large, undergraduate 'first
cycle', 'orienting' the suitable student towards a
more specialised 'second cycle' or letting him leave
with a broad qualification. The success of this will
depend on the means made available and the willingness of university teachers to change their ways.

Most of the other Savary reforms concern democ-
ratisation of the governing councils of higher

education establishments and attempts to stimulate
links between the universities and the 'outside
world'. One thing Savary does not do is end the
division of higher education into two separate
systems - the university faculties (open entry) and
the *grandes écoles*, with a competitive entrance
exam, preparation for which will have taken a year
in special preparatory classes. The most famous of
these *grandes écoles* are the *Ecole Nationale
d'Administration* (ENA) from which the civil service
high-flyers are recruited, Polytechnique which
produces the top technical and engineering-oriented
administrators and, incidentally, about half of all
the heads of large firms in France, and the *Ecole
Normal Supérieure* (literary studies). There are,
however, dozens of others providing vocational
training for the public service to a rigorously
selected clientele in every field from statistics to
public health, the judiciary, and of course the
armed forces, as well as the business schools.

Government

France since 1958 has been a very effectively
governed country. It has strong centralised executive power that is expected to be interventionist,
and there has been no particular demand to 'roll
back the state'. It also has all the normal attrib-
utes of a liberal democracy. The citizen can choose
his leaders at free and regularly held elections. He
is allowed to oppose the government and to join
organised groups to promote his interests. He is
protected by the rule of law, though some aspects of
the French judicial system give rise to occasional
doubts. Though the record of the state in relation
to television and radio is far from perfect, the
citizen can be said to have access to a variety of
information and opinions. There are the watchdog
institutions, not always entirely effective perhaps,
of a Constitutional Council and a parliament. Unlike
other Western democracies, however, France has had
executive power able to impose its will. It has been
able to pursue clear objectives - industrial modernisation, reduced energy dependence, grandiose
diplomacy - without encountering undue resistance
from industry, from unions, from environmentalists,
from parliament, from public opinion, from political
parties. In short, French democracy has lacked one
fundamental element that in the Anglo-Saxon world
contains the essence of the word democracy: the
mechanisms of restraint on executive power called

'checks and balances'.

The secret of this 'absolutist democracy' is the power of the President of the Republic. Provided there is a parliamentary majority prepared to support the Prime Minister and to pass the laws, implementing presidential policies that he asks it to pass, the President enjoys more concentrated power than any other Western leader. Like an American President, he is head of the executive and has the legitimacy of direct election; he has the prestige of being Head of State and Chief of the Armed Forces; he has great powers of patronage in the public service. An American President, however, has his appointments scrutinised at public Senate hearings and sometimes quashed, often has great difficulty persuading Congress to vote the laws and money he needs to pursue to policies, can have his actions investigated by congressional committees or the courts, has to campaign hard even for readoption as presidential candidate by his own party, has continually to defend his actions in public at press conferences which have none of the respectful character they assume in France, and can be impeached for 'grave misdemeanours'. No president in the Fifth Republic has had to put up with anything like that. A British Prime Minister, like the French President, can normally count on a parliamentary majority not to overthrow the government or frustrate its laws or policies. But the French President never has to answer parliamentary questions, make speeches in parliamentary debates on important occasions, or face parliament with statements on crises, international or domestic, and answer questions on them. In practice, presidential power in France is almost immune from the 'checks and balances' that control the executive in other leading democracies. The accountability of the French President is limited to a presidential election at the end of seven years - the longest term of office of any democratic political leader. An aura which can only be described as regal, the absence of direct accountability, and an increasingly wide range of policies and decisions which come within his ambit - these have been the characteristics of presidential supremacy in the Fifth Republic.

The French state has always been *dirigiste* and, when political circumstances have permitted a concentration of executive power, all-pervading. If one walks down one of the magnificent Second Empire boulevards in Paris, one should remember that each property developer received detailed drawings for the facades of every building from the offices of Haussmann's Prefecture. The events of May 1968 could be interpreted in part as an explosion of frustration, by students at any rate, at the paternalist state. Since then things have re-established themselves much as before. There has been no demand to 'roll back the state' that could be compared, for instance, with the prevailing 'privatisation' ideology of the victorious Conservative Party in Great Britain. The opposition parties in France have, it is true, promised to denationalise what the Socialist government nationalised in 1982. There are occasional suggestions from opposition figures and even from Socialists on the Social Democrat wing of the party that it may not be possible to go on financing the entire demand for health and welfare from tax and compulsory contributions. These are minor tremors. Furthermore, there has not been any great demand for more municipal or regional self-government. The decentralisation laws were willed from the top not from any groundswell of public opinion. The apparent exception to this is the increasing violence of Corsican separatists but this is nothing like so broad a movement for regional self-government as that represented by the Scottish National Party. Environmentalist pressure groups have nothing like the following of the Greens in West Germany. The Greens, like CND in Great Britain, are also a peace movement and these have no significant parallel in France. Even the Communists do not campaign for nuclear disarmament in France.

Yet the French state is less omnipotent today. This is attributable in part to a deliberately less absolutist style on the part of President Mitterrand and his government. It is also attributable to a fall in the rate of economic growth, which has put pressure on resources. Finally, the omnipotence of the executive is affected by the political situation since 1981.

The End of Absolutism

The Presidency. The new and less absolutist style of government is a deliberate aim of François Mitterrand, who belongs more in the Republican than the Socialist tradition in France. He was a parliamentarian and local councillor, representing the Nièvre in Burgundy virtually without a break from 1946 to 1981. He served as a minister in many Fourth Republic governments from 1947 to 1958. He condemned the return to power of General de Gaulle in

1958 because of the *"armed coup and sedition"* that brought it about: *"the presence of General de Gaulle signifies, despite himself, that from now on violent minorities can, with impunity and success, make war on democracy"*. He called the Fifth Republic system of government, in the title of a book he published in 1965, a *coup d'état permanent*. He was unhappy about direct election of the president because *"there is no example in France of a man swept to power by the direct confidence of the people and endowed with wide authority, without the rights and liberties of the citizens being in the end seriously reduced"*. When he himself first stood as a presidential candidate in 1965, he repeated the warning. He continued to be intensely critical of the concentration of power in the hands of the Presidency, denouncing 'L'Etat Giscard' where he saw executive choices worked out and decided in the Elysée Palace, parliament's role reduced to 'decrepitude', political interference with justice in all scandals which could touch those in power, all important public appointments, particularly those in the mass media awarded to 'the President's men'.

Mitterrand was never a member of the Socialist Party until he became the leader of the relaunched PS in 1971 and began to rebuild it as a credible alternative to Gaullist government, leading a united left alliance of Socialists, Communists and Radicals. He owes little to Marx and a great deal to the ancient tradition of the Republican Left, using words like justice, generosity, responsibility and liberty to define what he means by Socialism. Above all, however, the central theme of Mitterrand's political discourse is the old Republican opposition to 'personal power'.

Of course, as President he has a role to fill: he is expected by the nation that elected him to provide leadership. Presidential policy still sets the agenda for government ministers. Some important things have changed, however. There has, so far at any rate, been no instance of executive interference with the courts. This happened frequently enough under previous Presidents for a senior judge to write a stinging series of articles condemning lack of judicial independence, and for another commentator to say that *"justice has appeared in perfect submission to the solicitations of those in power"*. The real test of the Socialist government's commitment to judicial independence will be to see whether it fulfils its promise to end the present role of the executive in judicial appointments and promotions

through a *Conseil Supérieur de la Magistrature*
appointed and presided over by the President of the
Republic.

The Media. The second important change of
emphasis involves the mass media. Radio and television in France have long been considered far too
subservient to those in power. Under Presidents de
Gaulle and Pompidou, except for a liberal period in
Pompidou's first years from 1969 to 1972, the main
content and treatment of news and current affairs
programmes was more or less directly controlled by
the Ministry of Information. Under President Giscard
d'Estaing the emphasis slipped away from direct
control to control by patronage. The top appointments in television and radio news - even on the
commercial radio stations broadcasting to French
audiences from transmitters just outside the front-
iers (Europe 1, Radio Luxembourg and Radio Monte
Carlo) went to supporters of the President, often
to people who had served in his personal entourage.

Since the election of Mitterrand there has been
an apparent attempt to detach the media from government control, but at the same time a somewhat cont-
radictory, if understandable, removal of systematic
supporters and placemen of the previous government
from positions of responsibility in broadcasting.
A High Authority for Audio-visual Communications has
been appointed (on the same basis as the Constitutional Council: three members nominated by the
President of the Republic, three by the President of
the National Assembly, three by the President of the
Senate). It is this High Authority that will hence-
forth regulate broadcasting, make top appointments,
supervise political balance, and so on. It will
issue licenses to independent local radio stations
which are now permitted to flourish, though not to
finance themselves by broadcasting commercials.
Whether it will prove itself to be independent of
executive power and deaf to its wishes remains to be
seen. One of the problems of French political culture is that no one believes such a thing is possible,
though one does notice a greater readiness than in
the past to broadcast news items unfavourable to the
government in power.

Decentralisation. The third instance of a less
absolutist style of government is to be found in the
decentralisation laws, the first of which was introduced to parliament in July 1981 less than two
months after the new President took office. A strong
body of opinion had long held the view that the
French system of administration was too centralised.

Decentralisation was therefore to be the *grande affaire du septennat.*

Central control of local administration has been assured mainly through the Prefect, an official appointed by the central government who acts like a colonial governor in every department (roughly equivalent to an English county). He exercised *tutelle* - a word combining elements of control and supervision - over the communes and their local councils. In the past, all council decisions had to be approved by the Prefect. In recent years this had been relaxed somewhat, though he still had to approve the Council's budget and retained the power to annul decisions. He also acted as the executive authority for departmental councils and regional councils, the latter being non-elected and largely consultative bodies. The other strongly centralising factor was that practically every demand for a grant or loan for a development had to go out to one of the ministries in Paris for approval. It would be wrong to think that local councils had no scope for initiative, however. The *notables* who perform the prominent elected roles - mayor of a town or president of a departmental council - are important local and often national personalities. Such leaders might match, even outrank the Prefect under whose *tutelle* they officially came (the mayor of Lille, for instance, is the Prime Minister of France, Pierre Mauroy) and would have influence in the ministries to get what they wanted. The desirability for a town of having a mayor with influence in Paris serves, however, to udnerline how centralised this system was.

Under the reforms, the Prefect will lose his power of *tutelle* but remains the representative of the state in each department (with the additional title of *Commissaire de la République*). All council decisions and budgets will still be sent to him, but instead of the power to overrule he can now only refer decisions he considers illegal to the administrative courts or, for financial matters, to a regional Court of Accounts. He also ceases to be the executive authority for the department and the region. Some of his staff will be transferred to the direct authority of the presidents of the departmental councils. Nevertheless, as representative of the state, he represents all ministers (except Education which continues to be administered directly by the Ministry of Education and its regional officials); directs the services of central government in the area; and remains responsible for 'national interests, respect of the law, and public order'.

Another major change is the direct election of the 22 regional councils which will now have wider powers, including taxation. There have been regional councils since 1972, composed of the region's members of parliament and various representatives of local councils, but their powers were very limited. President Giscard d'Estaing, who in some fields tried, in his own words, for *"a more liberal function- ing of institutions"*, came out against regionalis- ation: *"I am entirely hostile to the political frag- mentation of France..."*

The importance of the new regional councils and the response of other councils to the reduced role of the Prefect will depend on how the transfer of functions and resources from central to local gover- nment works out. Certain functions, such as training and apprenticeship programmes, and the funds to pay for them, are to be transferred completely from the ministries to local and regional councils. For capital investment there is to be a Global Equipment Grant which will presumably be shared out among the communes by the regional or departmental councils, whose presidents will therefore be much more power- ful. It is much too early to assess these reforms. The habit of every local council of going to the Prefect or to Paris for subsidies and grants, and the habit of every ministry to control and direct the activities that come under it, are so deeply ingrained that the circuits may well re-establish themselves. Nevertheless, it is a bold step by the government. The Socialists in opposition often lamented the lack of 'counterweight powers' in France to stand up to the concentration of power in the centralised state. The lack of checks and bal- ances has been noted in this chapter. If the region- al and departmental elections, due to be held in 1984 and 1985, place the opposition in key positions through which they can determine the distribution of resources, the central state will have created a series of *contrepouvoirs* indeed.

We can therefore observe a deliberately less absolutist style of government - in respecting the independence of the judiciary, in at least reforming the direct intervention of the executive in the mass media, in pressing ahead with decentralisation. One needs to be a little further into the President's term of office to see if the good intentions remain when potential electoral disaster looms, and to see whether the more liberal institutions produce more

liberal effects. One must also remember that in the
economic sphere the state remains *dirigiste* - in
fact, through nationalisation even more so - in the
definition of objectives, the direction of investment, the handling of state-to-state deals and so
on. Furthermore, the Elysée is still in charge, and
the pervasive and occult influence of the Presidential and Prime Ministerial entourages remains a
feature of administration and political life.

The End of Omnipotence

The second explanation for the less omnipotent
French state of the 1980s is the one that occupied
the first half of this chapter - the effects of an
end to rapid economic growth. The real problem is
that as the rate of economic growth slackens,
demand increases for social transfers such as unemployment pay and social security. Add to that the
pressure of inflation on the costs of the whole
public sector and the rapidity with which a tolerable limit of taxation and compulsory contributions
is reached. Clearly, something has to be cut.
Unable to contain current expenditure and
interest charges on debt, the government cut back on
capital investment. Investments and capital transfers were 10.3 per cent of state expenditure in 1970,
6.9 per cent in 1980. Within this diminished total,
defence increased its share of capital expenditure
from 35.7 per cent in 1976 to 46.6 per cent in 1980,
while transport, communications, industry and
services were cut from 26.6 to 15.1 per cent.
Within what was spent on industry, a high proportion
went on sectors in decline such as steel and ship-
building. Investment in research was cut as a proportion of GNP - a process, it is true, reversed by
the present government. Of course this is all dep-
ressingly familiar to British readers, but it does
illustrate how it is growth itself that produces
modernisation, renewed investment, better roads,
trains, towns and, if that is what you want, prestige and *grandeur* as well.
The other blow to the omnipotent state dealt by
economic difficulties is the reappearance of the
veto power of interest groups. Interest groups are
always better at stopping something rather than
creating something. Nothing mobilises a group like a
threat to the short-term economic interests of its
members. However divided, weak and lacking in the
discipline of solidarity French interest groups may

be, whether employers or workers or anyone else,
they have shown themselves in the last two years
well able to prevent cuts in purchasing power (even
for objectives like a shorter working week) or
increased employer contributions for social security
and unemployment pay. While the economy was expanding rapidly, executive power was able to pursue bold
policy objectives, like the massive restructuring of
industry or the development of the hugh nuclear
power programme, without being blown off course by
interest groups. In this, France was different from
the more pluralist democracies like Great Britain or
America. With resources no longer expanding, interest groups cannot be pacified so easily.

The Fifth Republic in the 1980s

Finally, we come to the question of the Fifth
Republic itself. Amongst the problems that France
faces in the 1980s can we see the risk of a *crise
de régime*: one of those periodic upheavals which
from 1789 on have caused the collapse of whatever
system of government - monarchy, empire or Republic
- happened to be in place at the time? Some regimes
have been overthrown by revolution, some have ended
when the country has been defeated in war, some have
collapsed after a *coup d'état*. Could there be, if
not a violent scenario reminiscent of the past, at
least a *remise en cause*, a calling into question, of
the legitimacy of present political institutions?
The reason one hears this question asked is that
expressions of discontent are so much more audible
in France today than for two decades past. The
Presidency itself, after the defeat of Giscard
d'Estaing, has become vulnerable. The victory of
François Mitterrand seemed at the time like the
moving of the unmoveable. The alternation of power
from the coalition that had held it since 1958 to
an opposition divided between Socialists and
Communists had come to seem impossible: it would
bring about crisis, it would bring about Communism.
In fact, the transition was very smooth, and no
policy, domestic or foreign, has taken a Communist
or pro-Soviet turn. Now, however, that the delights
of getting rid of the government have been rediscovered, they are likely to become habit-forming.
Who today expects François Mitterrand, or even a
Socialist successor, to win the next presidential
election?
The discontent and the vulnerability, of course,

stem from the declining economic prospects. The
opposition is proving itself more effective in
exploiting discontent. Composed of ex-Prime Ministers and ex-ministers, with their own networks of
contacts in the civil service and industry, confid-
ent of returning to power before too long, it is
better placed than the previous opposition, out of
power for a quarter of a century, composed in part
of Communists whom top civil servants would not even
receive and whose attitude of permanent condemnation
tended to be regarded as mere ritual. The opposition
is using parliament more effectively; it has the
support of many newspapers, which daily propound the
view that in two years since 1981 socialism has
completely ruined the country; its views are the
views of the respectable and the influential in
industry and the professions, not, as before, of
Communist trade-unionists and left-wing intellect-
uals. It has a ruthless and determined leader in
Jacques Chirac, former Prime Minister, with an
exceptional capacity to attract the attention of the
media. He never misses a chance to exploit discontents and fears, real or imaginary: crime, immigration, soft sentencing, unemployment, prices, the
supposed influence of Communists on 'the disorgan-
isation of our economy' and on defence policy, the
supposed threat to pensions through the introduction
of early retirement. He has questioned the legitimacy of the President and called for new parliamentary elections. He has been accused of seeking
'destabilisation' of the regime.
    The word destabilisation brings one to the
question of legitimacy. Systems of government have
in the past lacked the commitment of the French
population. The Fifth Republic is the first system
to enjoy that widespread public assent we call
legitimacy France has had in the last two hundred
years. Unlike in past regimes, there is today no
major social group - Roman Catholics or Communists,
workers or peasants - which wishes to overthrow the
Republic. There is no reason why they should: there
is, after all, no oppression. The main institutional
innovation of the Fifth Republic - a powerful and
directly-elected Presidency - is widely supported.
Opinion polls since 1945 have showed a desire for
direct presidential elections. The interest shown in
these exceeds all other elections, and it is clear
that in voting people feel they are choosing the
nation's political leader.
    As the early systems theorists, like Easton,
pointed out, the stability of a political system

depends on inputs of support at three levels. In
descending order of importance these are support for
the political community - in this case the French
nation; support for the regime - in this case the
institutions of the Fifth Republic; and support for
the government of the day. If support for the first
two is strong, a low level of support for the
government is no threat to stability. One sees no
weakening of support at the two critical levels of
political community (except possibly in the case of
a minority in Corsica) and political institutions of
a sort that would turn opposition, albeit vehement,
to the government into a threat to the stability of
the system itself.

The Fifth Republic has overcome the greatest
test of stability - the peaceful alternation of
power from government to opposition in 1981. There
are still two tests it has not undergone. One is
whether popular commitment to its institutions
would withstand real economic hard times. Is legitimacy accorded only to the Fifth Republic as long as
it is reasonably successful, or could it, like
British institutions in the 1930s, for example,
remain unaffected even by a severe economic crisis?
The institutions have at least shown they can
'stand' two million unemployed without serious unrest, a fact that would have greatly surprised
people twenty years ago. The other test is whether
the system of government would work if a parliamentary majority was elected which was opposed to the
President of the Republic. This was much discussed
in the 1970s, but since the Left never won an
election the problem did not arise. One of President
Mitterrand's first acts in 1981 was to dissolve
parliament and a huge pro-Mitterrand majority was
elected. One day, however, there is bound to be an
election at which the people feel it is time for a
change and the majority vote for the opposition
while the old President is still in office. It may
well happen in 1986. There is no reason why this
should lead to a constitutional crisis. The constitution has up to now permitted a 'presidentialist'
reading, but it just as easily authorises a 'parliamentary' reading in which the President is obliged
to take account of the electoral verdict by appoint-
ing a Prime Minister acceptable to the new majority.
The Prime Minister would take much more of the
initiative in policy-making, but the President would
still retain important prerogatives like vetoing top
appointments and dissolving parliament if and when

he felt it necessary. There is no reason why the system should not work.

Less absolutist in style, less omnipotent in practice, one wonders, though, whether the French state will again recover its assurance as director of change and promoter of economic miracles. This provokes another, last question. Given French history, if the state cannot generate economic innovation, who can? That is a problem which the French may find more difficult over the next twenty years than the traditionally more liberal societies like Great Britain or America where the idea that the state can define and impose the public interest has always been resisted. In terms of self-confidence and vitality, the French, from Colbert to General de Gaulle, have always done best in the age of the state; the Anglo-Saxons in the age of the person.

Chapter Four

GERMANY

Geoffrey Pridham

        Taking into account the inevitable element of
campaign rhetoric in the SPD's programme for the
1976 Bundestag election quoted below, there nevertheless were and have continued to be important indicators supporting the then Chancellor Schmidt's argument for a "Model Germany". These were first and
foremost economic ones - lower inflation rates than
elsewhere, a middling level of unemployment, continuing export success and a strong currency - and,
indeed, the whole political reputation of Helmut
Schmidt during his Chancellorship (1974-82) hinged on
his performance as an economic manager. *"We have
created a Germany, of which many are rightly proud
and which enjoys respect in West and East. Domestically there is freedom and solidarity for the individual citizen, and externally freedom and solidarity as
a partner: that is the essence of our concept in
moving towards the 1980s. Our country owes its high
position to the performance and the formative will of
its citizens. Our country owes its high position to
our successful policy of rapprochement, to our
exceptionally high economic performance capacity, to
our tightly knit system of social security, to our
policy of constant reforms - and to the fact that we
have put into practice our intentions about promoting
social solidarity and real freedom for the individual
person! Therefore, we occupy a top place in Europe
and the world. That should remain so."*
        And yet the social and political dimensions
were also important; the former was stated in the
programme, while the latter was implicit. Schmidt's
concentration on economic management was not just an
unavoidable response to the difficulties which have
afflicted the world economy since the recession of
the mid-1970s; it also reflected his thinking that
stable economic performance was the essential pre-

requisite for preventing political misadventures, a consideration that clearly owed much to Germany's particular historical background. This interlinkage between political, economic and social factors has continued to determine the nature of policy debate through the seventies and into the eighties, as in previous post-war decades. Subsequent discussion in this chapter will show that concerns for political stability could easily surface when the state had to confront new and sometimes unsettling economic and social problems.

By objective standards of performance, the Federal Republic has matched up well to such challenges; but when it is viewed less from a political-institutional and more from a political-sociological standpoint, some doubts may be expressed. These tend to relate to the theme of democracy and to focus on such questions as political tolerance and civil rights, the extent to which political and social conflict may be absorbed (for a tendency to over-react is more present here than in most other West European societies), and an almost instinctive preference for seeking solutions to problems in a highly legalistic way (as one British political scientist wrote: *"the German tradition requires all things to be made quite explicit; it hankers after definitional exactitude – this constraint on politics is still in evidence"*). In retrospect, some of the shine has been taken of "Model Germany" if one considers other indicators than the economic, though even with respect to the latter the record has in some respects ceased to be so impressive: a negative balance of payments, a growing public debt and a rising level of unemployment during the 1980s, with little prospect of reducing this during the current decade. Financial restraints have more and more entered policy considerations in contrast with the reform momentum of the early 1970s.

Much depends on one's perspective in assessing developments in West Germany. Significantly, the self-congratulatory presentation of "Model Germany" was made in comparison with other West European states. In many of these, the Federal Republic is regarded with some envy as a stable and effective political system and a healthy economy. This favourable opinion abroad is important to the Germans, who are sensitive to the need for international acceptance and praise. Only on the political Left in certain countries – such as Italy, France and the Netherlands – is West Germany regarded critically, usually over certain issues of civil rights such as the *Berufsverbot* and response to terrorism.

From an internal, rather than external, angle, perceptions of the 'Model' differ however. This is no surprise considering the strong critical self-observation among Germans. While this may appear a valuable antidote to potential political disaster, an important part of post-war political behaviour (it was certainly lacking during the Weimar Republic), this mentality may be taken to extremes and can induce elements of political insecurity. Most notably, it has arisen over any appearance of extremism from either Right or Left. A widespread tendency to 'over-worry' has inevitably conditioned the positions adopted by policy-makers. This may be seen, for instance, in Schmidt's deliberately calm handling of terrorist crises such as the Schleyer Affair as well as in his determined emphasis on containing inflation, not least because of historical memories. But the habit of over-concern is not merely confined to the popular level: an instinctive anxiety about the 'constitutional order' being under threat sometimes appears among political elites - by no means entirely on the Right - at times of political and social tension. This critical self-awareness is one basic feature which differentiates Germany from other West European countries, in degree at least, if not in quality. It is distinctly less pronounced in Italy, the other main country which underwent the cauterising experience of Fascism.

Another standpoint is relevant to the discussion of policy matters. This is historical in the sense that political, social and economic developments since the late 1960s have not only introduced many unprecedented issues, but have also produced a shift in the nature of political debate.

The 'change of power' in 1969, with the appointment of a left-liberal coalition after two decades of Christian Democratic rule, undoubtedly contributed to the changed atmosphere of German politics. This event, linked as it was with such controversial policies as *Ostpolitik* (policy with regard to the East), promoted a polarisation between rival party elites which had wider repercussions in furthering politicisation at a mass level during the 1970s and after. What has above all marked off this third and now fourth decade in the development of the Federal Republic from the preceding two was the way in which political behaviour broke free from the post-war attitudinal straight-jacket imposed by the widespread rejection of 'politics' after Nazism and the almost

total devotion to the pursuit of materialism. This
had been encouraged by the restrictive atmosphere of
the Cold War and was exhibited in the authoritarian
style of leadership of Chancellor Adenauer.
     The 1968 movement was the first significant
break in this post-war pattern; but while it was
primarily confined to certain student circles, new
demands for participation among groups of society
outside political elites and activists have persisted
since then. This change has engendered a much less
deferential attitude towards the authority of the
state, though without it become anti-system. A major
consequence has been a more sophisticated awareness
of issues among the public. At the same time, tradit-
ionally-determined attitudes have remained and were
especially illustrated by reactions to law and order
issues, notably terrorism. It was significant, however, that the second alternation in government power,
in 1982, was much less fraught with trauma than that
of 1969, even though some policy changes were likely:
this suggested a growth in democratic maturity in the
intervening period.
     The enhanced mass interest in politics, as
evident in participatory demands and issue-conscious-
ness, had its effects in promoting social awareness.
But it is equally arguable that the latter nourished
the former. Such a development featured most strongly
in the awareness of the 'quality of life' which became visible in the early 1970s and grew apace thenceforth. It was above all instrumental in the attention
paid to environmental concerns, channelled societally
through citizens' action groups (*Bürgerinitiative*)
and politically through the 'Green lists'. This movement was significant because it expressed the emergence of post-materialist values, particularly among
the younger middle classes, which represented a turn-
about from the dominant line of thinking after the
war. The increasing inclination to operate outside
established political structures, i.e. the main political parties, was illustrated further by the anti-
nuclear energy protest movement (stronger in the
Federal Republic than elsewhere in West Europe) and,
again in the early 1980s, by demonstrations by
squatters in derelict houses. The latter was important as it reflected the social pressures emanating
from unemployment and promoting disaffection among
younger people.
     On a more general level, the period since the
late 1960s has witnessed a continued growth of secu-
lar as against traditional conventions. Some of these

were promoted through legislation, as the liberalisation of divorce and abortion and women's rights, but any estimate of how far the SPD/FDP coalitions carried through the 'inner reforms' promised in 1969 and induced changes in social structure and social consciousness must be highly sceptical. An example of reforms falling well short of expectations was in education, even taking into account the important achievements of expansion, as will be seen later in this chapter. Political and institutional constraints (far greater than those facing President Mitterrand after his election in 1981) explain much of the failure of these reforms, but so do econonic constraints once limited resources began to make a strong impact from the mid-1970s (as, in fact, the French Socialist government found after a year in power).

Economically, there have also been substantial changes compared to the earlier post-war period. Above all, the slowing down of economic growth and the impact of the oil crisis and world-wide recession have meant that the 'economic miracle' - much vaunted in earlier years - can no longer be spoken of so confidently. The catch-phrase 'Model Germany' really had negative connotations in that it was intended as a tribute to the Federal Republic's impressive ability to stem the worst effects of the new and uncertain economic environment. Inflation rates have generally been contained, although they did reach record levels for post-war Germany. In particular, energy, so essential to economic growth, has since 1973 become a top priority among governmental concerns. Unemployment has risen in the 1970s and 1980s to new heights, not seen since the mid-1950s when such figures were more tolerable because economic prospects were made bright by the need for reconstruction and the demands for goods, factors no longer applicable.

These economic issues all had political consequences. Firstly, the changed economic situation in which Germany found itself not only had a dampening effect on policy reforms, but also on broader political attitudes. It was a major factor in deflating the reformist euphoria present at the time of the 'change of power' and continuing into the early seventies. Indeed, both the Schmidt and Kohl governemnts have been characterised by economic retrenchment. A deeper question is whether greater economic stringency might have affected public opinion about the performance of parliamentary democracy. Secondly, the social market economy, firmly established by the Christian Democrats after the war, has been subject to some modification with the introduction of planning methods since the late 1960s, following the Grand Coalitions's efforts to overcome the recession of 1966-67. Thirdly, the federal system has itself been affected by the growing federal control over the finances of the *Länder*. The reduced autonomy of the state governments has also been a feature of education policy since the late 1960s, and is reflected in the adoption of new procedures to handle terrorism during the same period. All these aspects - political traditions focussed on the acceptance of democracy, the market economy and the federal structure of the state - differentiate Germany from other states in West Europe; and all have had some influence in moulding policy processes and policy solutions.

Finally, it is important to note the considerable interlinkage between different policy fields. Some of this has already become evident - such as that between energy and inflation, between unemployment and social protest, between limited resources and the prospects for reforms. Others might be added: the interlinkage between education and employment, and the often tendentiously interpreted connection between social protest and possible sympathy for terrorism. Several of the issues, particularly the so-called 'new' ones, have been difficult to handle because they do not fit easily into traditional ideological patterns and may be divisive within political parties. It needs hardly to be said that there is more complexity hidden behind the efficient and successful face of 'Model Germany' than may at first be thought.

The Economy

As with other West European states, the twin problems of inflation and unemployment have dominated the agenda of Germany economic policy since the early 1970s. These problems have produced a qualitative change of level in economic performance. The main feature has been a slow-down in growth as shown in the following figures. GNP was 97.9% up on 1950 in 1959; 49.4% up on 1960 in 1969; 28.9% up on 1970 in 1979. Productivity increased by 63.4%, 48.5% and 34.4% over the same decades. The number of people employed in 1959 was 21.9% higher than in 1950; 0.6% higher in 1969 than in 1960; and 0.4% lower in 1979 than in 1970.

Germany has been able to contain the worst effects of the world recession partly through her ability to export, partly because of reasonably harmonious industrial relations, but also through a successfully co-ordinated policy of stabilisation adopted early by the government. Nevertheless, as other countries, it has not avoided the general situation of uncertainty which has come to envelop economic policy-making.

It is necessary briefly to outline these problems as seen from Bonn. With regard to inflation, the federal government managed from 1970 to 1983 to keep rates well below the 'critical' point of 10%, with the highest level reached immediately after the first oil crisis. The annual rates were as follows: 3.3, 5.3, 5.6, 7.0, 7.0, 6.0, 4.3, 3.7, 2.7, 4.1, 5.3, 4.5, 5.0, 3.9%. Germany has boasted the lowest inflation rate for any European Community country since 1973, taking each country's rate per year, with the single and insignificant exception of Luxembourg's figure for 1973. Externally, the obvious main cause of the rise in inflation has been the increasing cost of imported oil following the Middle East war of 1973, and again the 'second energy crisis' of 1979-80. The Federal Republic's dependence on oil has been her salient economic weakness - roughly half her energy consumption is oil - and the import bill has rocketed. As Helmut Schmidt noted in his government declaration of November 1980: *"The second oil price explosion since 1978 has left immense traces in our economy: growing costs and growing prices, a balance of payments deficit and rising unemployment"*. Domestically, the sources of inflation have primarily been due to labour costs, but an inflationary spiral has been prevented by the fact that there has been no uninhibited interaction between price rises and wage demands (as in the UK). To quote Schmidt once more: *"Not all our difficulties have come from outside, but they would be much smaller if the decisive problem had not come from outside"*.

Inflation does, of course, mean less wages and less secure jobs. Unemployment has reached new heights. It was 150,000 (0.7%) in 1970; rose to just over a million (between 4.5 and 4.7%) in 1975-77; dropped somewhat until 1981, when it climbed from 6.7% to 8.4% in 1982 and 9.9% (2,500,000) in 1983. The German levels were again substantially below those experienced by other EC states. Nevertheless, since the mid-1970s unemployment has become structural and not merely cyclical. This has reflected the slowing of economic growth, but also demographic change (with rising numbers entering the job market) as well as industrial rationalisation. These factors are either not easily soluble or are long-term. The consequence during the latter half of the 1970s was a sluggish reduction in unemployment: a decline of places in industry compensated for by a growth in the tertiary sector, but with the latter reaching a saturation level in its capacity to absorb. In particular, the German labour supply has been immobile since a large proportion of unemployed are unskilled, while there has remained a persistent demand for more skilled labour. A worrying new factor in the 1980s is structural unemployment among youth.

Inflation

What have the governments attempted to do, and how far have they been successful in confronting these two fundamental problems? Have they followed any particular strategy, or have the uncertainties of the new economic environment meant that their policies have zigzagged in response to differing short-term pressures? Such questions cannot entirely be ignored when assessing policy content.

In this light, two aspects merit brief attention as conditioning government responses. The first has to do with attitudes. It is neatly summarised in the comment of one foreign expert on the Germany economy that *"the most important consideration is that no Germany government can be seen to compromise with inflation; to do so would be to commit electoral suicide"*. Historical memories of inflation, with all its possible economic and political consequences, have remained such that this powerful factor - it may indeed be called an element of the West German political culture - acts as both a constraint and a stimulus for policy-makers. Schmidt's policy of stabilisation, which became a first priority following the energy crisis of 1973, therefore enjoyed popular understanding and approval because potentially conflicting interests felt equally threatened by inflation. Public opinion polls have regularly confirmed that the public expects its government above all to keep prices stable. Insistence on sound money has even been strong among young people, for whom historical memories can only be transmitted experience. The evidence of opinion surveys during the 1970s and 1980s nevertheless suggests a more sophisticated grasp of complexities of economic developments, including their international causes and repercussions, a positive trend which owes much to the now high level of political awareness among the public as well as to the serious and thorough treatment of issues by the mass media.

88

The second aspect relates to the instruments of policy-making. The scope for government action and the content of its policies are partly at least explained by the fact that these instruments are differentiated in nature. Cooperative dialogue and bargaining characterise economic policy-making because of this dispersal of authority, which is shared between various political agencies (in Bonn the Ministries of Economic Affairs and Finance, not forgetting the Chancellor's office, and then also the governments of the *Länder*) and economic agencies (notably the Bundesbank in Frankfurt, as well as the commercial banks which are a major channel of relations with industry - which itself in this market economy plays an important autonomous role in co-formulating policy). A special note should be made here that the Bundesbank, as an independent institution not bound by instructions from the government, has been an im portant source of strict monetary control since stability became a top priority in 1973. In practice, this dispersal of authority has been matched by regular institutionalised cooperation between political and economic leaders, particularly under Schmidt who provided a focal point as Chancellor. This diffuse location of authority originated in the post-Nazi distrust of state power in the economy; but difficulties in managing the German economy have since the 1967 Stability Act led to the introduction of new instruments for coordination and planning at the national level along Keynesian lines.

The general assumption that inflation in the 1970s and after showed an unusual resistance to national efforts to bring it under control does seem to have found an exception in the case of Germany. It is above all in this field that Bonn has offered something approaching a 'model' to other countries. Germany presented an early example of following a tight monetary policy at a time when other European countries were still following expansionist courses. In 1973 the federal government adopted a package of deflationary measures involving steep rises in direct and indirect taxation, heavy cuts in public spending and borrowing, and cuts in liquidity. The Bundesbank supported these measures by sharply tightening credit restrictions. There were, however, no direct limitations placed on price rises, for, in official parlance, all state intervention in the 'free market regulatory process of prices and wages' was opposed. Further stabilisation measures in the same vein followed, but initial uncertainties about what the oil crisis fully entailed and the need to absorb its

consequences meant that it took a couple of years
before the inflation rate was brought down.
     This restrictive monetary policy, which has
characterised government policy since, was assured
by Schmidt's own commitment to stabilisation and by
the concordant line followed by the Bundesbank,
although at various point there emerged some shades
of difference between certain SPD and FDP ministers
about the extent to which monetarist priorities
should be carried at the risk of increasing unemployment. Following the further oil price rises during
1979-80, tight money control was once more applied as
the inflation rate again began to rise. The government concentrated on trying to contain home-made
inflation rather than attempting to stop the price
rises caused by the higher cost of oil, confirming
official thinking that in a market economy such rises
will stimulate conservation and substitution processes necessary for absorbing the higher cost of oil.
In line with this policy, the Bundesbank set itself
the target of maintaining the increase in money
supply within a 5-8 per cent range. One significant
feature in establishing this policy was the openness
of trade union leaders to Schmidt's argument that
this further oil price increase amounted to a permanent resource transfer and that it could not be compensated for by higher wage demands. Indeed, an OECD
report issued in the summer of 1980 attributed West
Germany's successful economic performance largely to
responsible trade union behaviour.
     This point has also been observed by foreign
governments looking at Bonn's success in handling
inflationary threats, including the British Conservative government which came to power in 1979 determined to follow a strict monetarist path. The West
German version of a strict monetary policy, however,
has been backed by consensus among both sides of
industry, with ready and effectively conducted cons-
ultations, unlike the UK. Chancellor Schmidt even
denied that his policy was a monetarist one: *"We have
achieved our success with our stability policy since
the beginning of the oil crises by a healthy mixture
of various instruments which were used simultaneously"*.
He elaborated: *"You can not fight inflation just by
applying monetary means; you also need fiscal means -
on the tax side as well as on the expenditure side;
some governments even believe in instruments in the
field of incomes policy, though we have never indulged
in this"*. Schmidt added, moreover, that he saw
Germany's policy as somewhere between the monetarism
of Britain and the USA under Reagan, and the expans-

ionist course favoured by France under Mitterrand.
This suggested a functional rather than a doctrinal
approach to economic policy in Bonn, for even though
anti-inflation efforts by the government and the
Bundesbank implicitly accepted the likely risk of
higher unemployment, selective expanisonist policies
have also been pursued as part of a dual strategy.
The German economy has by and large remained relatively healthy, so that the worst effects of any
conflict of priorities between fighting inflation
and keeping down unemployment - notably where high
interest rates have a depressive impact on output
and employment - have been avoided, although the outlook for this continuing is not encouraging.

Unemployment

On the problem of unemployment, the Federal
Republic offers less of an original approach. Unemployment has since the mid-1970s reached unusually
high proportions and by the mid-1980s a record level.
Measures previously taken to counter it tended to be
reactive rather than directive, providing relief for
those out of work, vocational guidance and job place-
ment services, but there has gradually been more
effort to rely on different instruments for guiding
the labour market. This change of approach has underlined official recognition that free competition
alone cannot provide a socially satisfactory re-
adjustment on the labour market, especially now that
unemployment has become a structural problem.
As early as November 1973, the federal government took the first measure to direct the labour
market by issuing a 'recruitment ban' checking the
influx of foreign workers (*Gastarbeiter*) from non-EC
countries (EC exempted because of the provision for the
free movement of labour). This reduced foreign labour
by some 670,000 by 1978, from 11 to 9.3 per cent of
the total workforce. Other special measures which
have since followed involve a mixture of regional
job creation, state allowances for investment, vocational training programmes and special provision for
groups like the disabled.
Generally speaking, these measures had only a
limited success in reducing the total of unemployed.
The Ministry of Finance estimated, for instance,
that the measures adopted in 1974-75 resulted in the
creation of 235,000 jobs during the period 1974-76,
and that the investment programme of 1977-80 produced
67,000 new jobs per year. Expenditure on job creation
measures rose from DM 17m in 1972 to DM 1,147m in

1980. Nevertheless, these efforts did not as a rule create permanent jobs, for employers were usually only ready to take on extra labour so long as public money was available. There have been other, more basic limitations on the impact of government measures: in the German market economy firms have been less inclined to react to government offers (lower interest rates, tax relief and investment subsidies) and more inclined to react to the general economic prospects, specifically growth - which has slackened, as noted above. From early 1982, when the total of jobless began to rise dramatically, the limitations of such measures were highlighted and the need for more systematic measures was underlined. However, the package of job-creation measures introduced by Chancellor Schmidt in February 1982 was generally considered a cautious initiative; they aimed at encouraging investment in the private sector through public money, in the hope that the private sector would put up its own money, and also at relieving the situation for the young unskilled.

There is one particular aspect of employment policy for which West Germany is renowned, and that is a traditionally well-developed vocational training system. This is used both as a means of improving skills among the workforce and as a cushion against youth unemployment. In the context of greater unemployment in the 1970s, the system was strengthened, and as a result it has contained unemployment among the young. In 1976, a full-time vocational year was introduced to broaden the occupational training of school-leavers and to meet the need for additional places. The Training Places Promotion Act of 1976 suggested a levy on firms if insufficient additional training places were offered to meet the increase in young people seeking them. Between 1975-79, the total of apprenticeships available rose from 145,000 to 625,000. All the same, so far as the job market in general is concerned, the major problem is that the age groups seeking first employment are now larger. This demographic trend is only likely to level off and diminish after 1990. The result is that the 1980s are seeing unemployment, which so far has particularly hit so-called 'marginal groups' (older people, the handicapped and those without vocational training), seriously affecting 'key groups' and especially young people.

Current trends thus raise new problems. Schmidt's aim in his 1980 government declaration of achieving full employment - that is, keeping unemployment below half a million - remained unfulfilled during his term

of office. Two years later, after the change of power,
the new Chancellor, Helmut Kohl, gave top priority to
solving the economic crisis and in particular to
creating new jobs. In his government declaration of
October 1982, he commented bleakly: *"Our own problems
with regard to economic growth, unemployment and the
budget result primarily from the fact that the German
economy is simply no longer able to cope with the new
challenges posed by the economic situation in the
rest of the world"*. Unemployment, which was not a
serious political topic before the 1970s, is certain
to stay at the top of the economic policy agenda. It
may even become politically more sensitive if 'key
groups' are more affected, with possible social
pressures and tensions arising. As we have seen, the
influence of government employment policy has been
restricted by a variety of factors largely outside
its direct control, notwithstanding the success of
measures in limiting unemployment. The conclusion may
therefore be drawn that while Germany still appears
as one of Western Europe's healthiest economies, it
has faced and will continue to face new and unfavourable developments compared with earlier times.

Social Policy

Social policy is often a blanket term covering
a bundle of multifarious issues linked by the common
theme of social change and development, and how
governments may - or may not - seek to implement
reforms in social structure and relationships as well
as respond to pressing social needs. In a real sense,
there is a social dimension to almost any field of
domestic policy. It is therefore not surprising that
social policy readily lends itself to ideological
interpretation; indeed, the extent to which a political party or government gives it a priority implicitly, if not explicitly, reveals its ideological
position. Parties of the Left invariably place a
higher priority on social questions than those of the
Right, simply because their understanding of policy
content and its consequences is more comprehensive.

Education

These general remarks have a particular applic-
ability in the German context for a variety of
reasons. Firstly, taking the ideological factor,
there was a pronounced self-demarcation on the part
of the new SPD/FDP coalition in 1969 from the previous
two decades of Christian Democratic rule precisely in

the area of social reforms. The catchphrase used by
Brandt was 'the beginning of democracy', meaning that
democracy required a social dimension, while the
tendency in the CDU was to view it essentially in
terms of the constitutional order. Education policy
is a suitable area to focus on since it easily falls
into the habit of ideological dispute, as in other
West European countries. Brandt himself selected it
for special mention in his inaugural speech as
Chancellor, when he asserted that 'the school of the
nation is the school'.

Secondly, while social policy may be ideological
in motivation if not in content, for the same reason
political and institutional constraints may well be
the more noticeable. Political constraints usually
refer to coalitions and their need for compromise
between coalition partners. In the case of education
policy, coalitional constraints were less in evidence
in the 1970s because the SPD and FDP agreed on prior
attention to educational reform and on strengthening
the federal role. Institutional constraints were, on
the other hand, very much present because education
policy was constitutionally a matter for the *Länder*.
Different parties have been in power in different
Länder, with consequent regional diversity in education systems. Party-political differences and conflict have arisen not merely between *Länder*, but also
between the regional and national levels of government once the latter's powers were upgraded constitutionally and the SPD/FDP coalition set about trying
to implement change. As experience showed, it tended
to underrate the constitutional constraints on an
active federal role in education policy.

Thirdly, social policy depends on the availability of resources because reforms usually require
public expenditure. The provision of personnel,
buildings and other facilities is crucial in education. Education also stands out among fields of
social policies as involving long-term considerations
in investment and strategy. Demographic change is, of
course, one of these considerations. The important
introductory point to note is that educational reform
was embarked on after 1969 when the economic outlook
was favourable, and that one major factor behind the
evaporation of the reformist drive has been the
diminishing availability of resources.

Another reason for examining the field of education policy is historical. The system that had
existed in pre-Hitler days was restored in its
essentials after the war. No major change occurred
until the educational reform movement from the mid-

1960s provided an impetus which, combined with the
1968 student protest, kindled public debate and
forced policy-makers to take note. There was talk of
an 'educational catastrophe', underlined by the fact
that Germany was way down the list with regard to
educational expenditure despite its being one of the
wealthiest of the West European states. The German
authorities had become very conscious of this by the
end of the 1960s. Education reform also offers a
case-study of planning methods being adopted nationally for the first time, so that any assessment of
results must include some reference to instruments as
well as policy content. Attention here will therefore
focus on the reform policies of the SPD/FDP governments.

Reformist pressures began to yield fruit once
the coalition came to power in Bonn in 1969, although
the constitutional reform of that same year was based
on a decision by the preceding Grand Coalition of the
CDU/CSU and the SPD. That amendment to the Basic Law
allowed Bonn to push more effectively for change in
the education system, for the federal government was
now granted important powers with regard to in-firm
vocational training, framework legislation for higher
education, cooperation with the *Länder* on university
infrastructures, the promotion of scientific research
and the regulation of educational grants. Various new
bodies were established to institutionalise cooperation between regional and national levels, such as the
Planning Committee for University Building and the
Federal-Regional Commission for Educational Planning.
With representatives from Bonn and the regional governments, the latter was an example of the more
effective 'cooperative federalism' that now ensued,
compared with the purely advisory bodies that had
existed before. Cooperation also proved more effective, however, for financial reasons. Bonn came to
exert greater influence because it provided necessary
resources for educational expansion, especially university buildings, and understandably exploited this
lever for political purposes. This is not to say that
Bonn, represented by the newly created Ministry for
Education and Science (which in 1969 replaces the
previous weaker Ministry for Scientific Research),
was able to turn the political tables on the *Länder*,
for the relationship between the two levels has been
complex.

The distinguishing feature of the SPD/FDP education policy was its social purpose, thus departing
from traditional concepts of 'education' as uncon-
ditioned by social factors, of scholarship as somehow

divorced from practical concerns. The Ministry of
Education produced a report in the summer of 1970
which spelt out policy thinking. In his foreword,
Brandt specified education as a 'measure of the
condition of a society' and as a prerequisite for its
economic performance, that is in promoting skills for
the job market. A strong leitmotiv was to break down
the traditional segregation between academic and vocational education. The report stressed that, while
the two post-war decades had been devoted to material
reconstruction, the state should now make up for this
'socio-political neglect'. Hierarchical values in
society should be replaced by 'a new democratic understanding', so that emphasis was placed on equal
opportunity and a restructuring of the system to
promote greater autonomy within its different levels.
In general, an integrated approach should be adopted
towards policy-making.

These were ambitious aims. They certainly
expressed the current trend of educational thinking
and the somewhat euphoric state of public interest
in reform at the time. As in other issues in German
politics, new vogues may develop and become intensely
pursued. But what results have come from policy-makers
adopting this outlook? Clearly, it was ideological,
but how far did institutional and political constraints, as well as the availability of resources,
modify policies? First, however, the achievements
must be examined.

University reform focussed on 'democratising'
their internal structures and enlarging their numbers.
The Higher Education Framework Law of 1976 established
the lines along which the *Länder* were required to
reorganise the universities. It broke with the traditional domination of professors, replaced the Rector
by an elected university president, replaced faculties
by new 'subject areas' (*Fachbereiche*) which controlled
resources instead of professors, generally widening
participation in decision-making, and improved coordination in courses and teaching methods. Although an
important milestone in the history of university organisation, its effects,so far as can be judged by the
mid-1980s, have varied between universities and
especially between *Länder*. The same applies to the
idea, also included in the law, of integrating different institutions of higher education into 'comprehensive universities': these have mainly been introduced in SPD *Länder* like Hesse and North-Rhine Westphalia.

The policy of building new universities was
based on laws of 1969 and 1976 providing for cooper-

ation between the federal government and the *Länder*.
The earlier seventies proved to be the great period
of university expansion: from 1965-78 the number of
students rose from 384,000 to 945,900 and of teaching
staff from 36,600 to 78,200. In this process, the
federal role was strengthened by the lack of resources
among the *Länder*, especially the poorer ones. While
impressive, this expansion has been criticised as
more quantitative than qualitative, conducted too
hastily to allow for sufficient progress and the
implementation of course and structural reforms. Since
the mid-1970s, discussion of course content has continued with special attention paid to relevance for
employment prospects, while the govenrment pressed for
the 'opening-up' of universities by reducing entry
restrictions, this being made possible by an agreement with the *Länder* in 1977.

On schools, the twin aspects of structures and
expansion have also dominated policy considerations.
But here the federal role has been less in evidence,
so that changes depended more so on the cooperation
of the individual *Länder*. This problem has characterised the introduction of comprehensive schools
(*Gesamtschulen*), in which West Germany is behind many
other West European coutnries. Following the opening
of the first of these schools during 1967-70, the
issue has been controversial along party lines, with
the CDU attacking the *Gesamtschulen* for being
'utopian', 'socialist' and based on 'class struggle'.
Predictably, policy has varied, with the greatest
number established in SPD *Länder*, though CDU *Lander*
have differed in the extent of their ideological
opposition. In general, the comprehensive school
system has made slow though not always sure progress.

Otherwise, structural reforms have concentrated
on changes within the existing three-branch system of
secondary education of the *Realschulen*, *Hauptschulen*
and *Gymnasien* (grammar schools), the former two concerned more with practical education related to the
demands of commerce and industry. Here again, there
has been much party-political variation in policy
between the *Länder*, though some CDU governments have
also introduced important changes. The most notable
reform has been the higher classes of the grammar
schools, with syllabus more flexible according to
social requirements and the needs of individual pupils.
Expansion in secondary education has occurred, though
less dramatically than in the case of the universities.
There was a significant increase in the number of
teachers (from 391,000 to 506,500 during 1970-75),
while expenditure rose from DM 16bn to DM 42bn over

the decade from 1970 to 1980 (the increase should not be overrated since it includes inflation in costs).

Measures extending vocational in-firm training have already been mentioned in the discussion of unemployment. While the federal government is responsible for this, vocational schools (*Berufsschulen*) came under the *Länder*. It is worth again stressing the priority accorded vocational education by the SPD/FDP coalition. This was partly in response to changing birth rates and the need for skilled labour projected for the 1980s; but it was also a reflection of the view that the traditional segregation between academic and practical education disparaged the latter. Various measures were introduced promoting and upgrading vocational education, including reforms in course content and better training of teachers, but as their adoption depended on the *Länder* there was again a lack of uniformity. The federal government continued, nevertheless, to encourage planning, and from 1977 the minister concerned was required to produce an annual report on vocational training.

Taken together, the education reforms amounted to considerable change, at least on paper if not in practice. The motivation was partly ideological, due to the election in 1969 of a government heavily committed to educational reform, but it was also a response to pressure deriving from the underdevelopment of the education system in the Federal Republic. Something had to be done as Germany was way behind most other West European countries. Recognition of this permeated even circles of the CDU/CSU, which helps to explain why the *Länder* run by that party went along with proposals from Bonn within the newly-established framework of national education planning. All the same, actual achievements fell significantly short of expectations held during the 'reform euphoria' of the later 1960s and early 1970s. Disappointment was understandable considering how ambitious these expectations were, but it is instructive to consider the policy-making process as this affected the content of policy. There were broadly three factors of relevance.

Firstly, systematic national planning in education predictably ran into the rivalries of the federal structure. Education remains largely a regional concern. Since the constitutional reform of 1969, Bonn has had one political foot in the institutional door, but very much as a partner in policy-making alongside the *Länder* and not in any way their master. Pressures for closer coordination have also meant friction between them. In practice, much has depended on the willingness of individual *Länder* and their powerful Education Ministries to go along with the wishes of the federal government. It is hardly surprising that, in the light of experience since reform began, demands for a stronger role for Bonn in education policy-making have reasserted themselves. These, of course issued, most volubly from Social Democrats, while Christian Democrats back in power (e.g. Federal Minister for Education) have been more acquiescent in the predominant role of the *Länder*.

Secondly, public opinion has visibly cooled, if not reversed itself. Here is an example of where the trend of public thinking has been both a motivating force and, later, a disincentive for reform. The relationship between public opinion and political leaders is usually something of a two-way process, for politicians may well seek to exploit new moods or use them as an excuse for a change of policy stand. CDU leaders used the less progressively minded public to campaign against comprehensive schools. Such an instance occurred in North-Rhine Westphalia, where in 1978 pressure groups of parents and teachers (backed by the CDU and the Catholic Church) forced the SPD/ FDP *Land* government to abandon its plan for extending the comprehensive system, after a successful campaign to collect sufficient signatures to call a referendum. This reversal of opinion cannot be simply written off as part of a 'neo-conservative' trend, although in a sense that label is true. The reasons range from the sobering impact of the more stringent economic situation on interest in reform to a certain fatigue with reform experiments. The latter was evidently in reaction to the very intense engagement in educational reform debates in the earlier 1970s, but it was also based on disillusionment with the operation of the reformed system (including the complexity of new regulations), especially among parents.

Thirdly, educational reform virtually came to a halt from the mid-1970s with the slowing down of economic growth. Planning is always linked to the existence and the earmarking of public funds: the shortage of finance has constricted any chance for further reforms. The recession gave the *Land* finance ministers a greater power than before to check additional educational expenditure and CDU *Länder* an extra argument against extension of the comprehensive system.

As a result the élan has long since disappeared from educational reform. Schmidt's statement in his government declaration of 1980 that the coalition had helped to break down class barriers through its educ-

ation policy may be true, but otherwise the section
dealing with this field was sombre and unadventurous
in its tone. Similarly, the policy of the Kohl government has not been particularly innovative; indeed,
its education policy concerns have been overshadowed
by the requirements of financial stringency, as shown
by cuts in student grants. All that can be said is
that reform in education was conducted just in time
before the recession his the Federal Republic.

New Issues

By new issues is understood not simply issues
which are chronologically new, but those which pose
a qualitatively different set of problems. The 'New
Politics' - a term which came into fashion during the
1970s in Germany - is identified with post-materialist
values and the quality of life, and, in the case of
Germany, is distinguished from the 'Old Politics' of
the economy and national security.
It is sometimes said that such new policy concerns do not easily fit into the conventional ideological framework or partisan alignments. This is
true in Germany of the issues of environmental protection, especially when related to the use of nuclear energy, although on other new issues like abortion, divorce and women's equality there is a delin-
eation between progressive and conservative stand-
points. So far as these issues - and notably nuclear
energy - have been divisive within parties, this has
become far more apparent on the left of the political
spectrum within the SPD (but also within the FDP)
than within the CDU/CSU. This, of course, presented
governments during the SPD/FDP coalition period with
an added problem in their handling of such issues.
The other basic reason why the 'New Politics' is
qualitatively different and extra-demanding for political authority is that the salience of such issues
is linked to a new level of public consciousness and
mobilisation.
The post-materialist values that emerged in the
1970s were pronounced because they involved a reaction against the overwhelming emphasis on materialism
in the first two post-war decades so strong that it
might even be called anti-materialist. The public
acceptance of an involvement in new issues has in
general been deeply rooted. The re-emergence of economic concerns from the mid-1970s has overshadowed
these issues in terms of government interest but not
necessarily in terms of public involvement. This divergence of interest between political leaders and the

section of the public most oriented towards new
issues, i.e. younger-generation, educated middle
class - is an additional reason why disillusionment
with the established parties has grown since the
later 1970s. In short, persistent public concern with
the 'New Politics' derives from value changes in
society away from conventional patterns of life, a
greater awareness that democracy means more than constitutional stability, and above all greater interest
in politics as such and in forms of political participation.

The two issues selected here have been dominant
in engaging public attention and challenging to political leaders, but illustrate two different trends
in political behaviour and thinking. Environmental
protection, and with it the question of nuclear energy, brought to the fore increased public interest
in politics, while expressing a decline in deference
towards political authority. Law and order, particularly terrorism, on the other hand, revealed that
deference and other traditionally determined attitudes were still present, and shows how far perceptions of political stability still appear to depend on
issue performance by a government. The common denom-
inator of these two issues was that they did not
easily relate to 'conventional' politics, that mechanisms for responding to them proved wanting, and
that, in different ways, they presented complicated
challenges to political elites.

Environmental Protection

Because pollution and other environmental
hazards have since the early 1970s been a subject of
consistent and growing public disquiet, political
leaders have been concerned to act on these problems.
It is an interesting example of where opinion has
convinced, or forced, government to adopt new values
to the extent that it not only early on passed a
series of special 'green laws', but also introduced
an environmental dimension to legislation in other
fields. As the government's report on its activities
for 1979 commented, *"the introduction of principles
for the timely consideration of environmental effects,
in the case of all public measures decided by the
federal government, is largely concluded in the individual policy areas"*.

As an issue, environmental protection dates
from the very end of the 1960s, although as a problem
it can be traced back further. With growing industrialisation and use of technology in what is a densely

populated country, the accretion of waste and consumption of different forms of energy has gradually
become an aesthetic and health hazard. Already in
1964, the first enviornmental laws controlled air
pollution and restricted the use of detergents.
Publicity on the subject began to increase from that
time, and protests about the need for social reform
helped to lend this question extra momentum. It was
the FDP, in opposition during the Grand Coalition and
free to indulge in new policy concerns, which first
took up environmental protection. The SPD was not then
interested in the issue, being more concerned with the
protection of jobs, but in the 1969 coalition it did
not oppose FDP pressure on the matter. Genscher, as
Minister of the Interior, acquired new powers and
received the support of Chanellor Brandt, a politician
who showed genuine interest in longer-term and even
futuristic questions and made the 'quality of life'
a theme in his election campaign of 1972. Thenceforth,
the government became committed to a policy of environmental protection.

The basis of the government's policy was its
programme for environmental protection of October
1971. It established ten theses which are worth
quoting because they outlined official thinking and
provided a clear foretaste of subsequent legislation:
(1) Environmental policy should ensure a human and
healthy environment, protect it against the adverse
effects of encroachment by people and remove any
damage. (2) The cost of enviornmental disturbance are
to be met by those who cause it. (3) Environmental
protection is to be supported by measures of a financial and structural kind. (4) Technical progress
must be conducted in a way that spares the environment. (5) Environmental protection is a concern of
each citizen. (6) A specialist council for environmental questions will be called into existence.
(7) The investigation, development and recording of
data for environmental protection is a political task
of the first order. (8) Study courses on environmental
protection are to be set up at universities.
(9) Environmental protection requires the cooperation
of the Federation, the Länder, the municipalities,
science and industry. (10) Environmental protection
demands international cooperation. Three basic principles underlay the programme: the principle of fore-
sight, which sought to prevent the occurrence of
damaging effects on the environment; the principle of
origination, whereby those who caused any damage
should be held responsible (this became known as the
'polluter pays' principle); and the principle of

cooperation, aiming at early involvement of social forces in environmental planning and close cooperation between the authorities.

The programme, which amounted to a comprehensive approach to a newly perceived problem was essentially carried into effect. It is not necessary here to list the complete run of laws, but some deserve mention: Removal of Refuse (1972), Lead in Petrol (1972 and 1976), Detergents (1975), Protection against Air Traffic Noise (1971), and the Law on Fighting Environmental Criminality (1980) which introduced into the penal code a new section on 'punishable offences against the environment'. According to an official commentary on the last of these, written by the Federal Minister of Justice, *"the environmental polluter is not be be considered worse but also in no way better than an arsonist, a swindler or a thief"* - an example of how German authorities attempt to deal with problems in a very legalistic manner. An assessment of the government's achievements in this field by the mid-1980s would include air and water pollution being checked; industrial dust reduced; an amelioration of lead and sulphur pollution; extension of the recycling process for paper, lead, glass and some metals; the problem of waste disposal largely solved; and the adoption of environmental precautions by the chemical industry. This impressive list of measures has been accompanied by the establishment of information and monitoring agencies like the Federal Environmental Office (under the Ministry of the Interior) and the Council of Specialists for Environmental Questions.

Policy implementation seems to have been a straight-forward process. The reason was a combination of the commitment of the SPD/FDP leadership to environmental policy and their direction of investment to it, as well as consistently strong public interest in the matter. The latter was all the more effective because it voiced itself through the flourishing citizens' action groups, which aimed at a whole range of environmental questions, local and national, and which, in their structurally diffuse form, amounted to nothing less than a mass movement. From the later 1970s, these *Bürgerinitiativen* channelled themselves politically through the ecologist parties or 'Green lists', thus having all the more impact on established political elites through their frequent success in state and local elections. This popular pressure, and the fact that it was difficult to argue against environmental protection except in a self-interested way, persuaded industry to comply with the new regulations, quite apart from

the fact that German legalistic attitudes had their
own compulsion. The one obvious source of reservation
towards the new policy was likely to come from industrial interests, because environmentalist ideology
called into question faith in economic growth and
because the new laws imposed additional costs. The
government attempted to drive a middle course between
conflicting demands, though it was not always easy to
establish exactly where such a course should lie.

However, from the mid-1970s there was a slowing
down of major legislation, essentially because of the
economic recession. Helmut Schmidt, the then new
Chancellor, was less committed than Brandt to environmentalist arguments; the trade unions turned their
attention more actively to the problem of securing
jobs; and the question of economic growth received
the highest priority. An important meeting took place
in the summer of 1975 between the government and
leaders of industry, with whom Schmidt came to develop
close relations, and thereafter new laws on environmental protection were less stringent than those
passed in the earlier 1970s. Furthermore, while
Christian Democratic leaders in government have
insisted on continuity in this field, the incoming
Kohl govenrment in 1982 showed a noticeable lack of
commitment to any active pursuit of enviornmental
policies in contrast to the enthusiasm of the new
Brandt government in 1969. In recent years, the
environmental question has in fact tended to concentrate on the nuclear energy issue, which became
politicised and proved to be the thorniest of the
problems relating to environmental protection.

It was over the development of nuclear energy
that the conflict between economic interests and
environmental considerations came most sharply to the
fore, and where the goverrmment found itself at logger-
heads with the environmentalist lobby. Government
policy, developed in the radically new context of the
first oil crisis of 1973, was one of energy saving and
fuel switching. Bearing in mind that Germany's only
indigenous source of energy was coal (which has a
limited potential, requires subsidies and, in any
case, creates its own environmental problems),
attention has turned to alternative forms of energy,
of which the nuclear option is the most obvious. The
government line has been one of limited extension of
nuclear energy, combined with a pronounced emphasis
on safety precautions. Despite the rational arguments
in support of this policy and the caution surrounding
its adoption, the use of nuclear power has progressively become very controversial, so that the powerful

anti-nuclear lobby, together with the complex and lengthy procedure for approving new nuclear stations, led to a reduction in the government's original programme from the mid-1970s. The issue was further complicated because the SPD/FDP coalition parties were strongly divided within themselves. This has, however, been far less true of the CDU/FDP coalition formed in 1982.

The arguments against the government policy have been on the usual grounds concerning the radio-active hazards contained in nuclear waste and the even more horrifying risk of a nuclear plant accident, a possibility that received a fervent airing in Germany following the incident at Harrisburg, USA, in 1979. This affair shook confidence in official circles in Bonn and, of course, increased public doubts about the security of the nuclear energy exercise. Although no such incident has occurred on German soil, there was a near catastrophe in Hamburg in 1979, following the discovery of a cache of nerve gas in a disused chemicals factory. It is clear that the anti-nuclear movement has capitalised on deep emotional feelings on the subject for which 'Gorleben' and 'Brokdorf' (nuclear power plant sites in North Germany) have become symbols. This movement, stronger in the Federal Republic than in any other West European country, has derived momentum from the trend of popular support for the *Bürgerinitiativen*, a growth of anti-elitist attitudes combined with the politicisation of groups outside traditional party activists, and an emotionalism which, rationally or not, looks back to recent German history. The press, for its own commercial reasons, has tended to fan the German partiality for 'worrying' with such headlines as (on the dangers from chemicals): *"Help - We Are Committing Suicide!"*

The nuclear energy issue is a striking example of where considerable constraints operating on government leaders, not least on Helmut Schmidt whose personal commitment to the controlled evolution of this new energy source remained strong, though this may apply less to Helmut Kohl because his government is less divided on the matter. It stood out as exceptionally difficult to handle compared with other matters of environmental concern. But even here the reduced interest of governments since the mid-1970s has only helped to nourish the ecologist protest movement. The happy coincidence between government policy, public concerns and acquiescent economic interest groups has long since broken down. The result is that environmental protection has been substant-

ially down-played since the period of frenetic legislative activity in the first half of the 1970s,
although its achievements remain on the statute book
and environmental considerations have become an
established consideration in the formulation of domestic policies generally.

Law and Order

'Law and order' in a literal sense comprises a
bundle of separate though often related issues ranging
from straightforward crime to highly political problems like terrorism. In his government declaration of
1980, Chancellor Schmidt mentioned the following items
in his section devoted to 'Policy on Law and Internal
Security': new forms of criminality, including economic crime (also known as 'white-collar crime', e.g.
credit frauds and the embezzlement of public funds),
the question of drugs, the improvement of data protection, 'radicals' in the public service (the *Berufsverbot*), political extremism and the fight against
terrorism. As to criminality in general, there has
been an upward trend through the 1970s into the 1980s,
among younger people, involving foreigners living in
Germany, in 'white-collar crime' and in street violence. The common theme in the handling of these problems, including also the more political questions of
law and order, is the operation of the legal code and
the provision of machinery such as the courts and the
police.

Since the end of the 1960s, law and order as a
policy area has undergone a sea-change. This is not
primarily a result of rising crime as such, but the
extent to which certain problems have come to the
fore, above all terrorism. It also derives from the
way in which problems become political issues. The
theme of 'security' was given considerable prominence
by the CDU/CSU as governing party during the 1950s
and 1960s, perhaps understandably as post-war Germany
felt very self-conscious about the durability of its
new parliamentary democracy. As an electoral appeal,
'security' referred particularly to the external dim-
ensions; as an internal problem, 'security' meant
in those days the threat from extremism of the Left
and Right.

With the 1970s, two new developments occurred.
The public became more actively concerned about law
and order. This was partly due to certain forms of
violence growing and especially becoming more visible,
notably street demonstrations and clashes with the
police. These had started around 1967 with the

Germany

activities of the Extra-Parliamentary Opposition, and
reappeared with anti-nuclear demonstrations from the
later 1970s and squatters' protests in the early
1980s. From the early 1970s the rise in Germany's
general crime rate has been among the most spectacular in Western Europe, but what matters in politics
is how this is perceived. The mass circulation press,
above all Springer's *Bild-Zeitung*, has regularly
given crimes of violence sensational coverage and
called for the harsher treatment of offenders. In a
country like West Germany, with its underlying reflex
about any public violence, whether involving a real
threat to authority or not, such media coverage could
have a powerful effect. The second development during
the 1970s was the emergence of terrorism, seen as a
new issue: no longer just a sub-division of 'law and
order' but a special problem demanding new legal and
institutional procedures.

The rise of terrorism virtually coincided with
the turn of the decade, although its origins may be
traced back to disillusionment of the earlier terrorist leaders and their supportive groups with the New
Left movement and '1968'. They felt that neither constitutional activity, as through the SPD which had
now entered government, nor even radical protest
within the established system, brought any dividends
in terms of a fundamental change in society. They
drew instead on the concept of the 'armed struggle'
from the Third World, disgust with Germany's Nazi
past, and a total rejection of consumer society - in
this specially violent form the anti-materialist
theme surfaced again. They were direct-actionist. Terrorism falls into different stages during the course of
the 1970s. Phase one occurred during 1970-72, the
highpoint of the Baader-Meinhof group operations,
which included bank robberies, bombings and arson.
The second phase during 1974-75 concentrated on kidnapping, the instituting of 'people's prisons' and
the attack on the West German embassy in Stockholm,
and involved new groups like the '2nd June Movement'.
The third and most dramatic phase came during the
'year of terrorism' in 1977 with the assassinations
of three public figures: Buback, the Chief Federal
Prosecutor; Ponto, a banker; and Schleyer, head of
the Employers' Confederation. It is possible to talk
of a fourth phase from 1978 up to the present, during
which there has been no clear pattern except that
much tighter security precautions seem to have had
their preventive effect. One new feature has been an
emergence of terrorist activity from the neo-Nazi
Right, such as with the bombings at the Munich

Oktoberfest in 1980, as well as actions seeking to
capitalise on anti-American feeling.

Any discussion of the terrorist problem must
also include its political and psychological repercussions. Germany adapts less easily to disorder than
most other countries. Various revealing reactions
were exhibited. The Munich Olympic massacre of 1972
produced a painful outburst of national shame because
it took place on German soil (although it was Palestinian terrorists who murdered the Israeli team) and
particularly because the victims were Jewish. Otherwise, public responses have tended to dwell too
heavily on the performance of the authorities in
dealing with an incident, probably more so than would
normally be found in other West European countries.
Criticisms of a 'weak' state emerged after the Lorenz
affair in 1975, but the firm handling by the government of the Stockholm embassy attack shortly afterwards occasioned widespread approval. The height of
public involvement in a terrorist crisis came during
the extended Schleyer affair in the autumn of 1977
(his kidnapping continued for seven weeks until his
death), when public nervousness was only alleviated
after Schmidt's successful Entebbe-style freeing of
the airplane hostages at Mogadishu. The Chancellor
was lauded afterwards as a 'strong man' and an
'admired German'. The political consequences of this
influenced government policy in handling terrorism.

It is difficult, however, to speak of a 'policy'
as such in the sense of policy content. In view of
the popular reaction to terrorism, with its implications for perceptions of political stability, a
special onus was placed on instinctive and immediate
political leadership. During the various crises
government action was closely observed by the public,
which was fed extensive coverage by the media
(although in the weeks of the Schleyer affair the
media did observe a self-imposed restriction at the
request of the government). Schmidt portrayed rather
self-consciously, though correctly, the style of
handling terrorist crises by his government: *"The
coalition reacted in a composed and appropriate way
and without dangerous emotion to the provocation of
terrorism, as was necessary in order to guarantee
the protection of our citizens and of our democratic
institutions without circumscribing the structure of
the liberal constitutional state through legislative
or executive over-reactions"*. He attempted to culti-
vate an above-party stance, which was reinforced by
the inter-party consultation through the 'grand
crisis committee', in which not only the Chancellor

108

and government ministers but also opposition leaders
and *Land* government representative participated.
Generally, inter-party cooperation was enhanced by
terrorist crises, even though afterwards polemical
exchanges reappeared and the then CDU opposition
indulged its law-and-order leanings. The most common
theme, though this was not confined to the political
Right, was that terrorism involved a direct challenge
to the Federal Republic's constitutional order.
Understandably, with all these pressures, the government's skilful handling of terrorist outbreaks was
nothing less than crucial.

Secondly, the government responded with new
legislation. The novelty of terrorism as an issue
and its unexpected appearance in the early 1970s
meant there was something of a lacuna in the criminal law which had to be applied in combating it.
The first measure was passed late in 1971, allowing
for strict sentences against highjacking and kidnapping. Further special measures followed in 1972,
but the main corpus of anti-terrorist legislation
was approved during the period 1974-78. The new laws
included a tightening-up of criminal procedure to
expedite trials; made the 'formation of terrorist
associations' a criminal offence; placed imprisoned
terrorists in temporary isolation (the 'contact
ban'); and considerably facilitated police searches.
Some controversy ensued about how far these laws
might restrict civil rights, even though majority
public opinion tended to side with demands for
strengthening legal procedures and security facilities. In the calmer years since the Schleyer affair
there has surfaced some discussion of whether the
anti-terrorist laws should now be changed or whether
they should have been limited to a set period, but
none of the political parties has shown any real
interest in this.

Thirdly, Germany's internal security system has
been considerably strengthened as a consequence of
the rise of terrorism. Legislation alone was hardly
sufficient to cope with the challenges imposed by
this unprecedented form of violent activity, as the
current phrase 'more state, but hardly more security'
suggested. Much additional government expenditure
was made available to reinforce such institutions as
the Federal Criminal Office and the Federal Office
for the Protection of the Constitution. The Ministry
of the Interior has created a special anti-terrorist
department, an electronic data system has been
developed, and the Federal Border Guard has had its
internal security functions extended to anti-

terrorist duties; as a consequence *Land* autonomy has
in some measure been scaled down in the cause of
greater efficiency. Altogether, this amounted to an
elaborate and fairly sophisticated security system,
though its effectiveness depended as much on its
efficiency when faced by terrorist challenges as on
legal provisions. There have been some inept and
almost unbelievable cases of bungling by the German
police authorities. Such episodes, speedily exposed
in the popular press, have only helped to make the
law-and-order conscious German public more edgy.
    Fourthly, there developed a somewhat more en-
lightened approach towards the problem of terrorism
in the calmer years from 1977, associated with
Gerhart Baum, who was Federal Minister of the Interior
duing 1978-82. This emphasised understanding the
causes of terrorism, thus implying that it should not
be seen solely as a law-and-order issue. Baum insti-
tuted government-sponsored research into terrorism.
His other line was a publicity campaign against
terrorism, using the fruits of this research, with
the aim of informing the public about the background
and roots of terrorism rather than engaging in polemical propaganda. A special event in this approach
was the seven-hour public debate between Baum and
the former terrorist Horst Mahler (of the Baader-
Meinhof group), as part of his carefully orchestrated
strategy to induce terrorists back into society.
These efforts did not exactly win the favour of then
opposition politicians like Strauss and others on the
CDU Right, who viewed anything relating to terrorism
as exclusively a matter of law and order. However,
despite the Ministry of the Interior being held by
the right-wing Bavarian Christian Democrat Friedrich
Zimmermann since 1982, there has been no new departure in anti-terrorism policy, perhaps because no
serious new outbreaks have occurred, though the work
of arresting terrorist leaders has continued.
    There have consequently been some signs since
the terrorist scares of 1977 that attitudes have
become more relaxed, though this clearly has much to
do with the fact that there have been no major crises
during recent times. Judging by surface evidence about
preventive action by the police, there have been some
possible narrow escapes, so that any prediction about
the future of terrorism as a problem remains speculative. Certainly, the elaborate articulation of the
internal security system has reduced the likely
incidence of terrorist activity, but it is less
certain that the roots of terrorism have been erased.
Disillusionment with West German society has, if

anything, grown since the later 1970s, although it
does not follow that this will necessarily turn to
direct action. One of the most interesting aspects
in examining German terrorism is the light it has
thrown on political attitudes. Here there are contra-
dictory signs. On the one hand, reactions to the
crude campaign against terrorist 'sympathisers'
during 1977 and criticism of the anti-terrorist laws
has sharpened public consciousness about civil rights,
especially among the younger generation already
deeply concerned about the illiberal implications of
the *Berufsverbot*. On the other hand, terrorism brought
to the surface a pronounced law-and-order mentality
among the public at large. It is probably an over-
statement to say that this revealed an authoritarian
potential in the Federal Republic in line with earlier trends in German history, for there is much evidence of changes in German political culture in recent
decades, but Germany did show a greater aversion to
disorder than other West European countries.

Conclusion

The period since the late 1960s has for a
variety of reasons - political, social and economic -
witnessed important changes in West Germany. These
have been most marked in the nature of political
debate, interest and invovlement, in the performance
of the economy and in the extent to which social
conflict has increased. Inevitably, all these changes
have been reflected in, if they have not helped to
condition, the policies which have dominated the
scene during this time. The case-studies discussed
above are fairly representative of the problems which
have faced governments in the Federal Republic.
What have these selected policy case-studies
got in common, and what have they illustrated about
the West German body politic? The tightening-up of
resources has placed a damper on reformist policies,
which were given an overriding priority in the domestic field by the incoming SPD/FDP coalition in
1969. Education and environmental policies exempli-
fied this problem, although significant achievements
were recorded in both during the earlier part of the
1970s. In the second half of that decade and into the
1980s, the government's priority switched decisively
to the performance of the economy, a change of course
not primarily due to the different policy interests
of Chancellor Schmidt compared with Chancellor Brandt,
but to the fundamentally transformed economic situation following the first energy crisis of 1973. This

priority has not essentially changed with the election
of Chancellor Kohl in 1982. At a deeper level, the
slowing down of the West German economy has apparently had differing effects. On the pessimistic side, it
relates to the growth of social pressures, and has
been most visible among younger people with restricted
employment possibilities and the lack of any encouraging outlook for the future. It has also been indir-
ectly evident in the greater gulf between political
leaders, with their overriding concern with economic
growth, and those social groups which question this
value, as on the issue of nuclear energy. On the
optimistic side, there is evidence to show that the
public as a whole has developed more absorptive
capacity in the face of economic severity, certainly
more so than was the case during the earlier recession
of 1966-67.

Another development illustrated by these policy
studies is that the role of the federal government
has grown. While state governments continue to determine important policy areas, their autonomy has been
modified in a variety of ways and this has been reinforced through institutional reform. In general, the
interventionist potential of the government in Bonn
has been enlarged through new mechanisms of 'cooperative federalism', as indicated by the growth of
economic planning, education policy-making and internal security systems to fight terrorism.

A further point is that the quality of political
issues has changed. This is not so much a question of
changing priorities as that many issues which have
emerged since the 1960s have been exceptionally difficult to handle and often complex in their subject
matter. This is particularly true of the 'post-
industrial' or 'new' issues. The other reason for the
change in the quality of political issues is that
greater political awareness and issue consciousness
among the public have placed extra demands on political leaders, and consequently have made policy-
making that much more exacting.

Finally, what is special about the Federal
Republic of Germany when looking at these policy
issues, all of which have in differing ways concerned
other countries in Western Europe? Certain established features like the social market economy and
the federal system obviously differentiate Germany
from other countries, even though both have undergone some alteration since the end of the 1960s.
Other particular features have emerged like a strong
awareness of the need for international cooperation
in a variety of policy fields, a pronounced emphasis

on legalism in solving political problems and, underlying this all, a persistent concern with stability,
both political and economic. This last factor touches
on what is probably the most distinguishing characteristic of West Germany and its politics, namely the
nature of political attitudes and their refusal to
take democracy for granted. This is understandable in
the light of history, but it has persisted despite
stabilising influences in the evolving political
culture and has affected the growth of mass interest
in politics. At the same time, while a positive
factor, this attitude has also expressed itself less
constructively as part of an habitual *Angst* syndrome.

Chapter Five

ITALY

P.F. Furlong

Differing Strategies

Other Western European countries may provide
models of decline, adaptation or renewal; Italy is
often seen as merely a particular case, an oddity,
the question-mark at the end of the Western European
sentence. Even Italy's successes, such as the
economic growth of the immediate post-war period,
are seen in these terms. A miracle, after all, is not
repeatable to order and is essentially inexplicable.
But of course that period of sustained economic
development known as 'the Italian miracle' is
certainly explicable in terms of the conditions and
policies of the time - a large reservoir of cheap
mobile labour, a strong stable currency, encouragement to political and economic integration in
Western Europe, a dynamic and expanding public
sector, among other factors. This pattern of economic
development provides, for many within the Christian
Democratic Party at least, the model to which policy
in the eighties should attempt to steer Italy. But
one of the major difficulties in understanding
policy responses to the crises of the seventies and
the eighties is that there is no consensus among
Italian policy-makers about 'the Italian model' or,
to put it less obliquely, about what constitutes
success.
How Italy has responded to the economic crises of
the seventies and eighties reflects not only the
objective factors in the international and domestic
environment, but also the differing intentions of the
various policy actors. The variations between these
intentions may often be superficially masked by an
overt commitment to the same set of words, when the
participants actually attach different meanings to
the generalities involved. To those who criticise

this policy style - an unstable compound of verbal
activism, practical incoherence and missed targets -
it must be said that despite the obscurantism and
evasion entailed, it is unlikely that there could
be even the small amount of coordination and cooperation which does occur without such sleights-of-hand
in the fragmented pluralism of the Italian system.
In the abstract and idealised Italian policy style,
politicians and administrators make considerable
efforts to frame policies in terms of their basic
values, perhaps because of a felt need to explain
their proposals to the sub-cultures in which Italian
politics is often held to be structured. In practice,
of course, the result is frequently a policy proposal phrased in impenetrable political jargon, which
might make one wonder precisely who is the target
audience for such prose. But the generalities and
the obscurities also provide scope for useful equiv-
ocation, which allows politicians to overcome a
crisis without losing too much ground within their
own constituencies, and the detrimental effects are
not felt immediately. The price of agreement is
incomprehension now, impracticality later.

The Christian Democrat Approach: Moderate Realism

There are at least three main approaches to the
problem, as it is often seen, of overcoming the
crisis and renewing the system. First, one of the
most influential is that to which we have already
referred. The Christian Democratic Party, in government since June 1945, has the difficult task of
affirming itself as central to the functioning of
the state while at the same time denying that it is
predominantly responsible for the present condition
of its charge. This problem is compounded by the
fact that whether the state depends on the DC or
not, the DC certainly needs the state to maintain
its own position at the centre of the sytem of
spoils allocation which has developed in the 1970s
and whose growth is largely responsible for the
apparently uncontrollable rise in public expenditure.
The DC, therefore, usually has little positive
incentive to reform the state in any way which
would radically reduce its capacity to use public
employment and social security as politically-
discriminatory welfare systems. Commitment to a
mixed economy with a large public sector and state-
controlled welfare is congruent with the traditional
populist Catholicism of the DC and with its inter-
class centrist identity, though Catholic social

policy would usually prefer the family or community to fulfil welfare fucntions where possible. Granted, also, that the DC is a coalition of factions in which ideological differences play little part but in which the maintenance of a secure position in power is of fundamental importance, it is difficult to find within these traditional elements of the DC any commitment to precise policies. Their language is that of short-term crises and long-term values.

The DC's electoral programme of 1979 performed the balancing-act with long-practiced skill: *"Two generations of Italians, with a continual effort aimed at liberty and justice, in the spirited dialectic between cultural traditions, social forces and political forces - among them and in the forefront the DC - have consolidated our democracy, and have brought about a profound social and economic transformation. Economic development and this extraordinary growth of liberty have resolved centuries-old problems but have also generated new elements of imbalance, which have been accentuated by indiscriminate attack against the Christian Democrat Party over the last decade. The result has been persistent and serious damage to the economy and institutions of the country. The interaction between these internal events and the prolonged difficulties in the international economy has contributed to the emergence of the conditions of crisis which we are now experiencing."*

Here one is given clearly to understand that the causes of the crisis are outside the control of the DC, and indeed that attacks on the DC serve to exacerbate it. The crisis is predominantly an international one, in which Italy is particularly vulnerable, partly because of structural imbalances but also because of the greater power of organised labour since the 'hot autumn' of 1969. The latter is considered to have increased labour costs by high wage demands above the rate of productivity gains, to have reduced labour mobility within the workplace, to have introduced rigid job security, particularly in the larger firms, and to have shifted income distribution away from profits and investment towards the wage-earning sector. The inevitable result, according to this analysis, is loss of competitiveness, price inflation, declining investment and recurrent balance-of-payments crises.

The fundamental precept behind the policies of the period of rapid growth (approximately 1953-1963) was international integration, at least of the most advanced sectors of the economy, which was facilit-

ated by liberal fiscal and monetary policy in the
1950s. This integration, and the early development
of certain sectors to which it led, was associated
with a high degree of specialisation in the Italian
economy around traditional, labour-intensive manufacturing industries, highly dependent on imported
raw materials and fuel. The structure of industry
which emerged from the so-called 'miracle' was
therefore one in which a high level of output was
tied to increased imports; the system worked, in so
far as it did work, because cheap labour costs and
high unemployment ensured a low level of domestic
demand. The tight monetary policy associated with
the political strategy of the centrist coalitions of
the period was an important factor in maintaining
these relationships.

The traditionalists in the DC do not imagine
that it is possible to return to those golden days
of growth without inflation; nevertheless, their
policies are based on the rolling-back of the gains
made by labour since 1969 and the use of unemployment once again to discipline in the work-force.
But the DC has restricted space to manoeuvre; it can
not pursue the rigorous deflationary policies acceptable to orthodox conservatives in Western Europe,
because it is not an orthodox conservative party. It
is difficult to think of many other major conservative parties in Western Europe which could have
found room in their electoral manifestoes for statements like the following, taken again from the DC
electoral programme of 1979: *"The experience of the
post-war period has shown that it is not the immis-
eration of the masses which changes the power-
relationships among groups and social classes; it is
the free development of the productive forces which
produces the growth in the power of the people
(strati popolari) and of the groups which represent
them. The DC has led this evolution and, as a great
popular force itself, holds it to be a fundamental
factor in social economic and political growth..."*
For the DC, its 'popular' identity means that
not only must it recognise the value of the shift in
power which has occurred, it must also claim to have
led and organised this shift. Moreover, it has a
clientelist base in the public sector of industry
and in the large welfare army, which restricts its
capacity either to increase unemployment or to
reduce the real cost of transfer payments. Nevertheless, as well as being a party with a Catholic
tradition and a clientelist base, it is also the
party of the conservative electorate - the party,

therefore, of order, of anti-Communism and of
political and economic integration with the Western
industrialised nations. Subject also to dissent
within its own ranks, to pressures from the
employers' confederation *Confindustria*, from the
Bank of Italy, from international actors (International Monetary Fund as well as member states of
the European Community), the DC is not able now to
pursue any consistent strategy, either following its
traditional centrist inclinations or satisfying the
demands of its other identities. But at times the
clouds part and, because of weakness of coalition
partners and favourable economic circumstances, the
DC appears briefly to be pursuing that path from
which it is so easily deflected: a high level of
economic activity with moderate wage increases
fuelled by increasing productivity; the unions are
kept in check by fears of unemployment together with
the consensual mediating powers of DC politicians;
a moderate level of inflation is tolerated, and
investment and productivity growth in the private
and public sectors ensure that international competitiveness is retained; overall, a buoyant economy
with a low level of internal consumption allows the
DC to maintain its clientelist welfare support,
without radical reform of the political and administrative structures which provide the channels for
this.

Of course, pursuit of this model makes Italy very
dependent on external political and economic factors,
and it must be acknowledged that the last time it
worked with even a minimum degree of consistency was
in the late 1960s, when the former official of the
Catholic-led unions, Carlo Donat-Cattin, was
Minister of Labour. It remains an important element
in any DC-led government as the 'best buy' policy
to which it will revert by preference. It reappeared
briefly, for instance, in the winter of 1975/76,
when its association with a very loose credit policy
resulted in a resurgence of inflation and a severe
exchange-rate crisis.

Though this 'best-buy' policy of the DC does not
represent a reversion to the 1950s, it shares certain
characteristics with the tradition established then,
in that it emphasises adaptation to the international
environment, it requires a low level of internal
demand, and it demands no planning of long-term
structural intervention by governmental authorities.
In this sense the traditions of pragmatism, of
modest objectives and limited horizons, which has
characterised Italy's foreign policy for much of the

post-war period, is continued, and is reflected in a
similarly unambitious approach to domestic issues.

Popular and paternalist, compassionate but firm,
centrist and radical, neither capitalist nor socialist, the DC still seems to see itself operating
within the context of the generic Catholic model of
the organic society, committed to a functional
individualism in which the freedom of each is found
in the full development of one's social faculties.
It is not integralist in its approach to Catholic
doctrine, nor is it merely the instrument of the
Catholic Church, but its model of society is akin to
Catholic corporatism, and its policy style reflects
the mediating consensual role which such a model
grants to its political representative.

Despite the flashes of Republican rhetoric, the
DC's tradition is the tradition of the commonplace,
of continuity and flexibility, a tradition which has
helped to cobble government alliances, for the most
part without firm policy commitments, for nearly
forty years. This tradition combines Catholic popul-
ism with the supposed demands of realism, which in
Christian Democrat terms means the recognition of
Italy's proper place as a stable and significant
part of the Western alliance, with the economic and
political limitations which that implies, but also
with at least outward respect for its position as
one of the leading industrialised nations in the
world, one of the big four of the EC, and an important element in NATO's southern flank. From this
come not only the openness of Italy's external trade
policy as a means of tying Italy into the West and
as a policy of advantage to the large multinational
industries of Northern Italy, but also the anti-
Communism. It is thus not only the DC's Catholicism
and economic policy tradition that militate against
the entry of the Communist Party (PCI) into the
government; it is also the notion that the inclusion
of the Communists might seriously question Italy's
international position for the first time since 1949,
not because of PCI policy, which is now firmly pro-
European, but because of the reaction of Italy's
NATO and EC partners.

This stance enables the DC to claim internal
successes as products of its guiding hand at the
centre of the state, and to avoid strategic choices
(such as the entry of the PCI into government) on
grounds of international limitations. This does not
mean that the DC is actively committed to a policy
of further European integration. On the contrary,
Italy's major governing party, though prone to talk

as though the development of strong supra-national
decision-making bodies is its constant aim, has
little incentive seriously to pursue such development because of the restrictions on its domestic
hegemony which would inevitably ensue. The lack of
obvious strong allies, both internally and externally, for a dynamic pro-integrationist policy is
also a deterrent to such a cautious party as the DC,
even if it saw an interest in such a development.

The 'Extreme Realist' Strategy

But an effect of the changes in the international economy since 1969 has been that even this
approach, which for want of a better phrase we may
term 'moderate realism', is now insufficient to
maintain Italy's economic and political position,
and Italian policy-makers now find themselves
increasingly under pressure to adopt a position of
'extreme realism'. This more radical strategy begins
from the assumption that, granted Italy's character
as a transformative economy, strong international
demand and a steady currency are essential pre-
requisites of sustained economic growth. So dependent is Italy on the actions of others in international affairs that the primary aim in the recession
should be to achieve a position to take advantage of
a higher level of world economic activity. This
could be achieved through currency competitiveness,
higher industrial productivity to be reached through
greater labour mobility, control of the wage-price
spiral, and the restriction of the Public Sector
Borrowing Requirement to allow room for private
investment to increase. The problem, according to
this analysis, is inflation at above the level of
Italy's competitors (16 per cent per annum in the
decade 1973-1983) and declining productivity.
The strategy would require a two-pronged attack
on the power of the unions and on the wasteful
inefficiency of the clientelistic practices of the
governing parties. Specific targets would be the
scala mobile ('the escalator', i.e. the system of
automatic index-lined wage increases), the complicated and wasteful state pension schemes, the government subsidy of temporarily-redundant workers known
as the *cassa integrazione*, government subsidies to
the public services, and the reform of public administration. In an influential interview during the
1983 electoral campaign, Guido Carli, former
governor of the Bank of Italy and president of
Confindustria, attributed the massive increase in

the state borrowing requirement wholly to three
factors: subsidies enabling the public to pay less
than cost price for public services, subsidies to
cover the losses of public-sector companies, and
state contributions to cover the deficits of social
security and national insurance agencies. He
continued, *"(This means) that the accounts of the
traditional state, of the state which provides
security, education and justice have not exploded.
It is all the rest that has got out of control. It
is not true that we are condemned to ever-increasing
deficits. We must only examine the possibility of
getting back onto a path of greater austerity, of
greater financial sobriety. At present we are
permitting ourselves as a collectivity things which
we cannot permit ourselves. That is all. There is no
mystery about it."*

A prerequisite for this strategy would be governmental stability, though precisely how this is to be
achieved is not clear. This strategy usually envisages some institutional reform also, with quicker
parliamentary procedures, greater governmental
control of the legislative timetable and possibly
the introduction of a constructive vote of no-
confidence. It is therefore of some significance
that this latter proposal was adopted by the DC
secretary Ciriaco de Mita in the 1983 election campaign. Another option for this strategy would be to
attempt to evade direct party responsibility in
government by the establishment of a so-called
'government of technicians'. The problem then would
be assuring party support for such a non-partisan
executive, but this obstacle, though serious, is not
totally insurmountable. Though the technocrats are
not dominant in any one party except perhaps in the
Republican Party, there are few experts in Italian
senior management, whether in the public or in the
private sector, who are not associated in some way
with one of the political parties - but then, of
course, direct party control of their supposedly
non-partisan nominee experts in the government would
defeat the purpose of the exercise. The 'government
of technicians' also has its attractions for those
who wish to see the PCI integrated into the executive without the problems that direct Communist
responsibility would cause for Italy's Western
allies.

Italy's international position, in this analysis,
is seen to be one of chronic dependence from which
Italy can only be rescued by her own efforts. The
stance of senior spokesmen for the private sector

has on occasions been markedly less complacent about European integration than that of the DC. Thus in 1976 Guido Carli wrote: *"The Community has shown itself unable to deal with the world's economic crisis and with that of the member most seriously affected - Italy...The apparent conclusion is that Italy will have to fend largely for itself in the period ahead."* There is no suggestion here that Italy can or should cut its links with the West. Carli's line is rather that Italy should adapt its domestic policy to its international limitations, whatever they may be. But private-sector businessmen are also increasingly impatient with the passive, low-profile foreign policy traditionally adopted by Italian governments. Just as in internal affairs the demands of political development required the granting of autonomy to important economic actors, so in foreign affairs the domestic multinationals were able to develop their own international financial and commercial links and to protect their own sectional interests abroad with some degree of independence. No radical choices were required of successive Italian governments, the choice of trading, military and political partners having already been fixed in the post-war settlement.

In the difficult economic climate of the 1970s, the problems of declining productivity of the Italian work-force and of increasing foreign indebtedness restricted the scope of the large-scale Italian multinationals, whose directors began to take a new interest in the lack of foreign policy and to require a more active role from Italian governments, both in energy policy and in the development of new commercial links outside the ambit of Italy's usual trading partners. However, it is important to note that such demands do not usually envisage Italy as a bridge between the EC and the Third World, or as the natural independent leader in the Mediterranean. The extreme realist position does not wish to endanger Italy's traditional links with the industrialised countries of the West, but it is not clear how these are to be preserved while Italy stretches to reach Southern Mediterranean or Latin American markets and resources. On this point the extreme realist tends to lapse into generalities. Thus in 1976 the Chairman of FIAT, Gianni Agnelli, argued that the increasing fluidity of the international system allowed countries more room to manoeuvre, and continued:*"The probable evolution of the international situation will impose a front-line position on Italy...The alternatives facing Italy are to continue with a low*

*profile policy without making clear choices in inter-
national affairs, in which case Italy will inevitably become more and more an object instead of a
subject of international politics; or to accept real-
istically what the situation calls for - greater
international involvement - and to try and transform
into strengths the very factors which are thrusting
Italy onto the world stage. However, for this to be
achieved there will have to be imagination and
creative leadership."*

Spokesmen for the extreme realist position are
not as numerous as those for the moderate position,
but their offices and status are significant. In the
Bank of Italy and among the large-scale private
industrialists, as well as among Italy's international creditors, such views seem to be increasingly
popular. For much of the post-war period, this
important sector of the Italian economy has had no
firm political base, having initially been linked
with the Liberal Party and later with the Republican
Party as well as with some factions of the DC. In
1976 it appeared that the main representative body
of Italian private industry, *Confindustria*, was
hopeful that the prospect of the entry of the PCI
into governing circles might sharpen the appetite of
the DC for the radical reforms which the strategy of
extreme realism demanded, either because some of
them were also acceptable (for different reasons) to
the PCI or because the DC could not risk cooperation
with the Communists without paying some ransom to
the right. In fact, the DC could have given partial
satisfaction to both the PCI and *Confindustria*
(which is not as difficult as it might sound), but
failed on both counts.

With the failure of the historic compromise
either to achieve governmental stability or to rope
in the industrial working-class with the promise of
future benefits, the *Confindustriali* have finally
begun to turn their attention towards direct
involvement at the centre of the political system
through the DC. If this alliance were to come about,
it would mend a long-standing rupture between the DC
and predominantly-lay or liberal private industry,
and would constitute a compromise no less historic
than that between DC and PCI, though one that would
invite less international attention. The election of
Carli as an independent senator with DC support in
1983 is a move in this direction. In the electoral
campaign, however, he proposed a temporary wealth
tax, which pleased the Socialist and Communists in
part but displeased the DC, and a one-year wages and

salaries freeze, which annoyed all three. In the
same campaign the DC secretary, De Mita, appeared to
be committing the party to a line of austerity and
modernisation close in sympathy to the extreme realist strategy put forward most cogently by the then
Governor of the Bank of Italy, Carlo Azeglio Ciampi,
in his 1983 annual report. Indeed, one influential
newspaper editor went so far as to say that there
were really only two electoral programmes, that of
Ciampi and that of Giorgio Ruffolo, the Socialist
economist. The electorate, as electorates are wont
to do, chose neither. The DC suffered its worst
electoral setback since the war, losing nearly 6 per
cent of its vote, and its brief convergence with the
Confindustriali was an immediate, though not necessarily permanent casualty.

The 'Alternative Strategy' of the Left

The flirtation with the Mediterranean temptation
which is apparent even in Gianni Agnelli's position
is more marked among the parties of the Left. If the
other perspectives consider themselves in some sense
realistic, that does not mean the strategy of the
Left should be categorised as idealistic or unrealistic. If a term is needed which would be acceptable
to the Left, one might describe it simply as 'alternative': in its assumptions it is certainly that,
even though in some policy proposals there is coincidence of views. Whereas the parties of the Centre-
Right and the private sector seek a combination of
institutional reform and reduction of government
involvement, in the view of the Left the problems
really begin with the failure of planned government
intervention in the 1960s. The latter have tended to
view economic programming as the only possible
solution to the chaos of Italy's large public-sector
operations. That does not necessarily entail further
nationalisations. On the contrary, the aim is initially to bring existing state intervention under
rational control.
    The need is therefore not to roll back the gains
made by the working class since 1969 (and thus
restore the supposed equilibria of the 1960s), but
rather to find a new international role for Italy
which takes account of domestic and international
changes. In this perspective the problem of inflation
is at least partly attributable to parasitic clientelism in the public-sector and to the wasteful
inefficiency of public expenditure, as well as to
international factors. But the major problem is not

inflation - it is employment, with internal regional imbalances and increasing international marginality close in importance. Without some state regulation, the low-grade technological industries of the Italian economy, which are small-scale and highly labour-intensive, are at the mercy of the large domestic and foreign conglomerates. Without some state intervention, advanced-technology industries cannot grow and there is no effective control of the input of primary resources on which a transformative economy such as Italy's is very dependent. In any case, the Italian economy is already characterised by massive state intervention which is accepted by all major economic actors - the aim is not to withdraw completely from market forces but to ensure that the public sector operates in the public interest and without waste. This, in the view of the PCI, can best be achieved by a serious policy of planned restructuring.

The priorities of the parties of the Left and for the unions, with of course some differences of emphasis between them, are the return to full employment, the removal of regional imbalances, particularly those affecting the South, the protection of the gains in conditions of employment made since 1968, and diversification to reduce Italy's international dependence. Inflation, to which the extreme realists attach almost mystical significance, is a problem for Italy only in so far as it does not affect Italy's competitors to the same extent, and the massive public expenditure deficits which Carli and Ciampi attributed to unproductive state subventions of various sectors in the economy are attributed by the Left to the excessively high rates of interest which the Bank of Italy pays for short-term loans to the state, as well as to necessary state support for employment in the public sector.

The solution to both these problems would be a cautious, state-directed reflation of the economy, through lowering of interest rates and planned industrial restructuring. The alternative strategy would also give higher priority than the realists to the increase of tax revenue. Italy appears to impose a relatively low fiscal burden on its citizens in terms of tax-take as a proportion of GNP, but this hides the extent to which revenue is drawn more efficiently from wage-earners (covered by a recently-introduced pay-as-you-earn system) than from the very large self-employed sector. In view of the high tax rates already applciable, the alternative strategy envisages improving revenue by more aggressive

Italy

pursuit of tax evaders, in particular illegal
exporters of capital.

The dynamism that the Left require of the
Italian state would also apply to foreign affairs,
and like the extreme realists, it would like to see
Italy more active in pursuing links with developing
countries. It is, of course, more prepared to
regard the EC as in need of radical reform, and has
a more instrumental approach to it than the DC. The
initial rejection of the PCI has now been substantially altered, and since 1972 the Communist representatives in the European parliament have been on the
whole constructive, seeing in a supra-national
European grouping a potential counterweight to the
power of the USA and USSR. International integration
is therefore no longer a matter of joining the 'rich
man's club', as it was once described, though that
is not to say that the Left is at all happy with the
limitations imposed by Italy's dependence of the
West and the USA in particular for commerce and
credit. In 1977 Francesco Forte, then a senior
spokesman for the Socialist Party and later a
Minister of Finance argued speculatively: *"For the
intermediate countries, which have only limited
power, free trade is of use to avoid their becoming
excessively dependent on one of the strong nations.
The principal reason why we are interested in
remaining in the European Community is perhaps to
avoid falling excessively into the orbit of the USA.
It is certain, however, that if Germany were to show
a lack of concern for us in terms of this sort of
solidarity, then they would push us financially and
therefore technologically towards the USA, and the
cause of a European Common Market of intermediate
economies could gain a lot of ground."*

This was said at a time when European cooperation was even more under threat than usual, and the
European Monetary System has since silenced some of
the voices in Italy which were prepared to doubt
German commitment to solidarity. But the EMS is not
redistributive in impact and has actually made
currency competitiveness more difficult to achieve
for Italy. For the Left at least, Italy's undoubted
commitment to European integration should not mean
a total commitment to an unreformed Community. In
practice, Italian monetary authorities have tended
to use devaluation as a compensation for domestic
inflation, despite the existence of the EMS, and
have fought hard to maintain the competitiveness of
the lira in the face of devaluations by other
member states, in particular France. Other major

Italy

sources of conflict are the Common Agricultural
Policy, which is seen as weighted against
'Mediterranean products' and as inappropriate to the
traditional pattern of Italian agriculture, and
industrial policy, where EC attempts to force Italy
to cut back its steel production meet sustained
opposition in the public sector. In these fields the
emphasis given by the Left to full support of
domestic capacity conflicts both with EC opinion and
with the other domestic political strategies.

The three main approaches to the management of
the crisis do not exhaust the variety of perspectives in influential sectors of policy-making, but
between them they cover the important themes in the
policy debates. In general, it might be said that
underlying the rhetoric there is a perhaps surprising degree of convergence about the need for reform
and about some of the measures necessary, even
though the values and assumptions underlying the
proposals are in conflict. The need for quicker
parliamentary procedures, for greater technical
expertise in administration, for the reduction of
overmanning in administration, for more effective
control of public spending - all of this is undisputed, and in Italian terms has the same substance
as a sermon in praise of virtue. The track records
of successive Italian governments give no grounds
for expectation that such reforms are near implementation or even near agreement.

New Participants

The problems to which Italian policy-makers
address themselves are not necessarily those faced
by other countries, nor are they addressed in
precisely the same manner. Among the reasons why
even the reforms accepted as necessary are not
treated with expedition is the fact that the stability of the state and the persistence of old bad
habits tend to deprive the policy debate of its
urgency. The debate is dominated, therefore, not by
the crisis of the moment or by specific, long-term
problem areas, but by the more generic need to
achieve a pattern in governmental activity that
might resemble normal functioning. The problems
posed by the energy crises, the low-growth economy
and the international recession have not by themselves stimulated any profoundly new approaches to
policy-making. Rather, the policy debate reflects
the failure to reach a standard of day-to-day
efficiency which does not involve periodic break-

down in a variety of sectors. The 'commonplace'
policy style ascribed above to the DC does not carry
with it technocratic or rationalistic impulses, and
this undoubtedly contributes to the failure; but the
DC does not face the electorate on its record of
administrative competence - this issue, though no
doubt important and certainly a source of concern
among educated elites, does not usually appear to be
of electoral salience.

The changes that have occurred, such as they
are, in the character of policy issues and in the
ways they are confronted have tended to result not
from the Christian Democrats, nor even from the
governmental apparatus; they have come from alternative elites, usually excluded wholly or partially
from the sphere of government, who on rare occasions
have forced the issues, and the reforms, on the
reluctant elites in power. This may be done through
temporary coalitions in parliament, through referendum appeals which compel parliament to take
action, through unusually effective pressure exerted
by coalition partners on the DC, or through rather
more covert interest-group action.

Secularisation and Individual Rights

Perhaps the major change in recent years in
Italian political culture, which has been concurrent
with the rise of alternative elites, has been the
decline in the capacity and willingness of the
Catholic Church to involve itself directly in party
politics through support of the DC. The secularisation of Italian society has proceeded in a dramatically rapid fashion, and it has proved possible for
relatively small groups of militant anti-clericalists
to introduce major changes in legislation such as the
Divorce Law (passed in 1970 and confirmed in a referendum in 1981). The functioning of the Abortion Law
has been hampered by the instruction given by the
Church to Catholic doctors and nurses not to cooperate, which appears to have had a wide effect.   The
new Code of Family Law promulgated in 1975 altered
radically the previous Code which had been based
largely on laws passed in the Fascist period. Changes
in education policy have affected the power of the
Church in schools, and the reform of health-care
provision appears to have transferred significant
control of the local welfare system away from relig-
ious orders towards party politicians or party-
affiliated administrators.

Italy

On the whole, the process of secularisation has
found the state unable to respond with reforms to
meet the social changes, and the running in this
field has been made perhaps more than in others by
active, hitherto-excluded groups. A would-be
umbrella organisation for such efforts was provided
for some time by the Radical Party which campaigned
on a wide range of civil rights issues. The relative
success of these campaigns in an otherwise immobile
state is due not only to the vulnerability of the DC
at a time when it also is undergoing some secularisation but also to the anachronistic nature of the
issues involved. Changes in divorce and abortion law
refer to an ideological conflict over the relationship between the state and the individual which in
many other European countries has been settled for
some time. Because the changes are typical of a
liberal state, they are less costly and less difficult to implement than changes relating more directly
to issues in the welfare state.

The Labour Movement

Two other major internal changes have occurred
which have susbtantially altered the pattern of
policy-making. The first dates from the large political mobilisation of the student population in 1968
and of the trade-union membership in 1969. Whereas
in France the events of May 1968 did not fundamentally alter the long-term distribution of power, in
Italy the mobilisation of the late 1960s, while
slower burning, had wider repercussions and its
effects, both political and economic, are still
being felt.
One political effect was to introduce new vocal
minorities into the political arena who demanded
greater participation in effective decision-making
and who had established a tradition of violent mass
protest. None of the major political groupings
proved capable of integrating the dissident student
groups, and from this tradition sprang not only the
left-wing terrorists of the 1970s, but also (from a
different element) the concern with direct democracy
and civil liberties which now characterises the radical area.
A rather different kind of political ripple was
produced by the trade-union movement, which derived
from the 'hot autumn' of 1969 a lasting stimulus to
joint action and a vigorous concern with industrial
and political reforms. If the student movement put
civil liberties and political toleration onto the

agenda, then the trades unions might be said to have
made themselves into an issue. Though their capacity
to influence political decisions has varied since
1969, there can be little doubt that Italian governments, which before then were able to exclude large
sectors of the conflict-ridden labour movement from
power, have since then been compelled to include the
trade union confederations in any distribution of
resources.

The ways of doing this have varied: up to 1972,
the after-effects of the 1969 mobilisation ensured
that governments were willing to listen to the joint
confederations, while the left-wing parties wanted
to win back the unions from the spontaneous industrial actions which threatened to detach them from
party control. Hence one of the first fruits of the
newly-developed union muscle was the passage in 1970
of the Workers' Charter which codified many of the
improvements achieved in a piecemeal fashion the
year before. Since 1972, the unions have combined
local industrial action with political pressure at
the national level in pursuit of a variety of
objectives. Their policy concerns have been mainly
the defence of, and at times improvement on, the
gains made between 1969 and 1972, particularly in
the fields of wage-indexation, labour mobility and
job security, but they have also attempted, with
less success, to put forward a version of the alternative economic strategy. The unions' major problem
is that without pressure from them such reforms are
very unlikely, but that union action is largely
ineffective and potentially counter-productive,
since it risks destabilising precarious government
coalitions.

The closeness of the union policy to that of the
Communist Party does not indicate dependence, and at
times the joint confederations are willing to make
agreements with employers or governments in the face
of PCI opposition. In March 1980 a spectacular
example was provided by the agreement between the
unions and the Forlani government over an austerity
package which included adjustment of wage-indexation
together with a novel 'solidarity fund', with union
participation in its management, to provide funds
for investment partly from industrial workers' sub-
scriptions. Strong PCI opposition and an adverse
reaction from the shop-floor defeated the *scala
mobile* side of the proposal. A more successful
example of a DC-led government's attempt to show
that it did not need PCI backing in its negotiations
with the joint confederations was the 1983 agreement

on the adjustment of wage-indexation. This episode
provided the unusual spectacle of a public clash
between Luciano Lama, secretary of the Communist-
dominated confederation, and Enrico Berlinguer, secretary of the PCI.

The Regions

Though the unions have emerged as national political actors in policy-making, their effective autonomy from the parties, and even more their capacity
to wrest changes in policy from the government, are
still very limited. The same might be said about the
other new group of actors in national policy, that
is the regions. The law instituting the ordinary
regions in accordance with articles 114-133 of the
Constitution was passed in 1968, and the first
regional elections were held in 1970.
For the Socialists, the new regions were intended to be an important instrument in the planning
process, while the DC viewed them with anticipation,
apprehension and antipathy in about equal parts. The
decisive factors in the implementation of this radical innovation had more to do with the short-term
party-political considerations and electoral pressures than with agreed views of policy needs within
the governmental coalition, though the fact that the
regional reform had undisputed constitutional authority was also important. Much earlier, a DC
Minister of the Interior had referred to the
Costituzione-trappola (the 'snare-Constitution') and
this aptly summarises the wary attitude adopted by
successive DC-led governments to some of the more
radical aspects of the founding document of the
Italian Republic.
The assumption in the 1960s was that the regions
would in some way implement regional plans in harm-
ony with national goals embodied in long-term plans
established by central government. Since such
economic planning has never functioned properly at
the national level, the integration of regions with
the policies of central government has not proceeded
in the way intended. On the contrary, the lack of
clear central direction, the emergence of claims by
the regions to participate in the formulation of the
national budget, and regional demands for more
decentralisation of government policy-making, led to
the blurring of the lines of demarcation of responsibility between regional and central government.
Since tax reforms gave central government increasing
control of the money notionally available to the

regions, a major effect of the regional reform has actually been to introduce new participants into the confused pluralism of Italian policy-making and to render conflict between the new participants and central government virtually inevitable.

The integration of this conflict, which might (in the intentions of its 1960s proponents) have been resolved by a national economic plan produced in consultation with the regions, is now achieved in an ad hoc fashion by the development of joint central-regional bodies which administer policies in a variety of sectors such as agriculture, industrial reorganisation and credit for small industries. There is no fully coordinated approach to regional administration by central government. On the other side, the Conference of Regional Presidents has recently emerged as an important national voice for elected regional officers, but even here the strength of regional party control ensures that unity of opinion, leave aside unity of action, are difficult to achieve and apparently impossible to maintain. The effect of the economic crises of the 1970s has been to make central-regional conflict more acute, particularly where local employment is affected by central policy, and to put more pressure on the control of the purse-strings exercised by central government. In the field of regional policy, as in others, fragmented pluralism magnifies conflict, delays decision-making and allows participants to exercise reciprocal vetoes.

As well as the differing policy objectives, so the changing instutitional structure and the vagaries of socio-political development have also affected the way policy issues are handled. These factors have determined how policy is made and what solutions are adopted. The rest of this chapter will consider some of the major issues that have appeared in Italy in common with other European countries in recent years, but it must already by evident that while international constraints are of crucial importance to Italian politics, that does not mean that the problems of adaptation to those constraints are necessarily the problems to which solutions are sought. In part this is because some of Italy's problems are specific to Italy, and relate not to the energy problem, international recession and the instability of international currency arrangements, but rather to Italy's separate development within Europe.

Italy

The Economy

For the purposes of explanation, it is easiest
to begin with similarities: Italy certainly shares
with other Western industrialised countries problems
of price inflation, and increasing unemployment; but
though growth rates do not match those of the 1950s
and 1960s, the economy retains an underlying buoyancy which shows itself in a relatively rapid pick-
up of demand once monetary pressures are relieved.
The problem is that Italy's public expenditure has
risen rapidly in the last decade, appears to be
uncontrollable, and consumes a large part of funds
notionally available for credit because of the large
deficits entailed. Because of Italy's dependence on
imported raw materials and on imported energy
resources, the balance-of-payments constraint also
looms large.

Inflation and Balance of Payments

Precisely where these issues appear on the political agenda in relation to one another depends on
whose agenda one is looking at. Carlo Ciampi referred to the lack of political will in the face of
inflation, *"...a call to reality, the re-affirmation
that a stable currency is a precondition of development and of ordered civil community"*. Italy has had
little success in bringing inflation down to the
level of its competitors, but evidently, for better
or worse, has learnt to live with it.

Part of the reason, of course, why price inflation has been tolerable is the system of wage-
indexation, which ensures that arguments over wage
increases are fought out at national level at three-
year intervals, when labour contracts are due for
renewal, rather than in a piecemeal, repetitive and
localised manner at factory level. The joint confederations have until recently been successful in
preserving the real value of the unit increase in
the *scala mobile* despite increasing opposition from
the employers. This covers mainly employees in
medium to large sized firms; other sectors of the
population are not so well protected in real wages,
but some are covered by the complex sickness and
retirement pensions schemes, whose rapid growth in
recent years helps explain the increase in the
public sector deficit. This is particularly so in
areas of high under-employment and unemployment such
as the hinterland of the South, where entire communities appear to maintain themselves on such benefits.

134

In other sectors, such as the small artisan-type
industries, these protections are not to readily
available, but it is precisely in such sectors that
moonlighting, untaxed employment in second and third
jobs, is prevalent. These mechanisms do not mean
that price inflation does not cause real hardship,
but they do help explain how relatively high levels
may be tolerated over considerable periods.

Though the social effects of inflation may be
mitigated in a variety of ways, the lack of external
competitiveness which it induces (because Italy's
competitors have lower rates of inflation) is less
easily circumvented. The fact that Italy's growth
rates have actually been stronger than most of its
competitors since the late 1970s has exacerbated
this problem, both because of Italy's dependence on
imported raw materials and because of the high
levels of internal demand in consumption. The deterioration in terms of trade which follows bouts of
sustained higher activity in the Italian economy
demands remedial measures from the Italian authorities, and one remedy chosen appears to be periodic
devaluation. For example, from March 1981 to March
1983 the lira was devalued within the EMS on four
occasions. The lira is nevertheless stronger than
this might suggest, since the last three were
clearly competitive, induced by changes in the rate
of exchange in France, whose economy is considered
by the Italian authorities to provide a benchmark
for Italian performance, and by the need to retain
export markets in West Germany, which is Italy's
largest single market abroad. Devaluations against
the Northern European countries are also a stimulus
to tourism from the area. Since Germany is the largest single source of imported manufactured goods,
such currency realignments have an inflationary
cost-push impact while at the same time serving to
dampen domestic demand.

The other major instrument used in an attempt to
achieve a sustainable equilibrium is deflation
through high interest rates and control of credit.
Here the close connection between employment policy,
industrial policy and general macro-economic management becomes clear. It has been argued that since
the oil crisis of 1973-74 the Italian authorities
have pursued with only limited success a restrictive
policy on internal demand, and have periodically had
to sharpen this (for example in 1977 and 1981) with
so-called 'austerity' policies consisting of emerg-
ency import controls, higher prices for public
services such as transport, electricity and gas, and

medical prescriptions. The purpose of these policies, according to this line of argument, is not solely to avoid hitting the balance of payments constraint but also to secure a redistribution of income from wages to profits, through inflation, higher unemployment, and stability or decline in real wages.

The cumulative effect of these policies, if successful, would be to restore domestic and external equilibria to something like their pre-1969 pattern, but the domestic equilibria would be illusory as they would actually consist of low growth and high unemployment. The persistently high unemployment levels since 1977 do seem to indicate that the authorities have more on their mind than just external balance. Whether this 'grand design' theory of macro-economic management is applicable or not, it does seem to be the case that the financial orthodoxy of the extreme realist position is more visible in the persistent deflationary policies of recent years than is the hopeful 'moderate realism' usually associated with the DC. But it is probably futile to attempt to discern a coherent policy giving unity to macro-economic management in recent years. Governments show inconsistent, cross-pressured reactions, and are subject to wide variations of approach over short periods of time, particularly when, as is usually the case, the economic ministers come from different parties.

Italy's problems are not only caused by high oil prices: the domestic causes of inflation and external imbalance may be equally important. The compensating mechanisms referred to earlier, such as wage-indexation and very high spending on welfare, also tend to build inflation into the system, even if they do not directly stimulate it, and may contribute to the 'crowding out' of private and public sector capital investment. When Italy does have current account surpluses, these have tended not to feed mainly into domestic demand; a good proportion has filtered back out of the domestic economy through net capital outflows, in part via illegal exits. It has been estimated that in the period 1961 to 1972, the average annual net capital outflow from Italy was 1.1 billion dollars, and in the period 1965-1972, 1.6 billion dollars. These capital outflows might be regarded as another compensating mechanism since they appear to be triggered by fears of inflation and of excessive Treasury spending. As a result partly of the IMF intervention in 1976, more attention began to be paid to this problem of illegal export of funds, but the effectiveness of

the measures taken is a matter of doubt, particularly since they depend to a considerable extent on
self-policing and policing of clients by the commercial banks.

Industrial Policy

Italy's industrial structure is characterised by
a variety of factors which are closely linked to the
occurrence of persistent inflation and periodic
balance-of-payments crises. The nature of Italy's
postwar economic development was such that the economy is now highly specialised in a relatively narrow
sector - footwear, textiles, domestic electrical
products and motors. In crucial sectors such as
steel, chemical products and food, Italy runs a
persistent deficit and is likely to remain in
deficit despite the recession. Another aspect of
this is that Italy specialises in sectors which are
vulnerable to competition from the Third World, and
its role in the world economy, such as it was, is
therefore directly under threat from that quarter
also. These problems were evident before the oil
crisis, and the dependence on external energy supplies which had been obvious before was one more
structural problem in an economy whose internal
contradictions, limitations and strengths were already established in 1973.
     But what is striking about the Italian experie-
nce of recent years is not so much the failure to
respond satisfactorily to the energy crisis but
rather the lack of a serious industrial policy,
certainly before the Industrial Restructuring and
Reconversion Law of 1977 and arguably even since
then. For much of the post-war period, 'industrial
policy' has been synonymous with 'Southern policy',
where the aim was to bring standards of living and
rates of growth in the South up to those elsewhere
in the country. Initially (up to 1959) this was to
be achieved through the use of the free market:
government funds were used in build up infrastructure in the hope of encouraging an industrial revolution which would then develop autonomously. From
1959 to 1971, the strategy involved direct intervention by the state in the building of large-scale
industrial projects, the so-called 'cathedrals in
the desert'. Since 1971, industrial development in
the South has been, at least formally, in the control of the regions, but there has been little central
coordination, and the easy access to credit which was
the government's intended main contribution has

become a maze of complicated, time-consuming procedures depending to an increasing extent on discretionary powers, therefore susceptible to clientelistic manipulation.

The 1977 law, while not simplifying matters, did
attempt to provide a national policy including the
large public-sector corporations and covering the
entire country, with intended emphasis on the development of high-technology industry and encouragement
to small business. How little it succeeded in its
aims of encouraging diversification and higher productivity in Italian industry may be gauged from the
fact that in 1983 the Minister for Industry declared
that his first priority was to rewrite the existing
laws on industrial policy into a unified and radically changed text. A major problem with the ambitious 1977 law was the failure of the appropriate
regional and national government bodies to produce
realistic development plans for the different sectors of industry. As a result, industrial development
funds tended to be spent in a haphazard and unproductive manner.

Other persistent problems in industrial policy
have been the opposition of the relatively efficient
and large-scale private-sector industries of the
North to government planning, and the failure of
governments fully to control what should have been
their most important instruments in industrial policy, namely the public corporations, IRI and ENI,
and the state-owned commercial banks. The reluctance
of private industry to cooperate with government is
partly the result of the historic division between
the liberal lay elites and the newer Catholic interests. Because of dislike of high public expenditure
implicit in the DC clientelist-welfare policies,
Italian entrepreneurs have tended to put their
money into speculative finance or capital export
rather than into constructive investment in industry.
The approach of the left-wing parties, combining
verbal aggression with practical acquiescence, has
also engendered hostility and misunderstanding in
the private sector. The outcome of this ambiguous
attitude of the parties of the left has been called
'government by juxtaposition' - over-rigorous legislation followed by facile suspensions and exceptions.

The Italian experience of state enterprise,
particularly through IRI, has been much analysed and
for a period in the 1960s was an object of admir-
ation for some foreign observers. Briefly, the
system of management by which it operates is based
on the principle that the achievement of social

objectives will involve costs which cannot be covered by prices established through market competition. The social costs are met by a holding company (IRI, in this case) from a state-provided endowment fund, while the productive subsidiaries retain their entrepreneurial role. The system requires careful monitoring of costs and clear specification of social objectives in advance by the political authorities. If these fail, the state-owned holding company risks becoming a collector of wrecks, and there is every indication that this has been happening to IRI.

Coordination within the public sector is supposed to be provided by the Minister of State Participation and by the Interministerial Committee for Industrial Policy. In practice, the most important power of the minister has been the appointment of directors, but he has had little control of the corporations' investment strategy, largely because of their reluctance to divulge financial information, but also because of the assumption of directive powers by the interministerial committee - on whose effectiveness doubt is cast, in turn, by the way ENI, the energy and petro-chemical holding company, recently disregarded directives regarding massive investments in the Sardinian operations of its loss-making mining subsidiary.

Employment

The recession took some time to affect employment in the traditional industrial areas. Though official unemployment has increased from about 1.1 million in 1971 to about 2.5 million in December 1982, the crude figures mask a complex pattern of growth, stagnation or decline in different sectors. In particular, three points should be noticed.

First, the bulk of this increase in unemployment occurred from 1980 onwards. It is only since then that the job rigidity bemoaned by the industrialists for much of the 1970s began to relinquish its grip, making way for managerial rationalisation and productivity increases particularly in the motor industry; the collapse of the FIAT strike in October 1980 marked a turning point in labour relations and the adoption of a new aggressive style of management, as well as allowing FIAT to proceed with mass redundancies in the main Turin factory. As so often in Italy, FIAT's lead was followed quickly in the private sector by other large firms such as Olivetti and Pirelli, and in the public sector by Alfa-Romeo

and Snia Viscosa, all of which have successfully
implemented plans to shed labour. It is not yet
clear whether the restructuring is including new
capital investment on a large scale.

Secondly, since industrial employment was well-
protected in the 1970s, it is not surprising that
the detailed statistics show that there was actually
a small increase in the numbers employed in industry
between 1971 and 1980 (from 7.61 to 7.7 million).
Employment in industry in the EC as a whole declined
by about 10 per cent in the same period, and the
only other EC countries to show an increase were the
very different economies of West Germany and Ireland.
This development against the trend shows up in other
ways, for instance in the much greater resistance to
the world slump in demand for steel: alone among
EEC countries, Italy increased its ferrous metal
production between 1979 and 1981, and throughout
1982 and the first half of 1983 continued to object
loud and long to EEC directives on reduction in this
sector. By 1981, however, employment in industry had
begun to decline as the recession, which previously
had been felt most strongly in agricultural employment and among those seeking first-time jobs, started to erode the traditional northern industrial base
also. The services sector continued to expand at
the expense of the other sectors, but not sufficiently to mop up the excess labour.

Finally, the true extent of unemployment is
masked by the incidence of short-time working, which
the OECD in December 1982 estimated to represent the
equivalent of 300,000 full-time jobs in Italy, and
by the persistence of the black economy, whose
effects are more difficult to quantify precisely.
Some indication of the size of the phenomenon may
be gleaned from the fact that in 1978, after much
political pressure, the government statistical
office finally adjusted its method of national
account aggregating to rectify the previous under-
estimation caused by refusal to acknowledge the
extent of the black economy. The sector most affected was firms with less than 20 employees, which in
the old series of accounts provided 32 per cent of
total value added in manufacturing industry in 1975,
in the new series it is nearly 60 per cent. The net
effect of this and other modifications was an 8 per
cent increase in GDP at current prices in 1975, new
series over old.

Italy

Public Expenditure

If Italy appears to have weathered the recession with relative success, the price of the apparent stability which is the net content of that success lies at least in part in the largest compensating mechanism of all - that is, the large increase in public expenditure since 1974. Though there are those, mainly 'extreme realists', who view this as a problem in itself because of their belief in control of the money supply, there is a widespread concern among monetarists and non-monetarists alike, for two main reasons. First, the expenditure increase has not been covered entirely by a corresponding rise in income, and, second, successive governments are having increasing difficulty in controlling the rise in public expenditure.

Though these problems were certainly visible in the 1970s, they began to dominate policy thinking and constrain policy measures severely in the period after the dissolution of the 'historic compromise' parliament in 1979. Between 1980 and 1983 the state's current revenue increased from about 38 per cent of GDP to about 44 per cent; in the same period, current public spending increased from about 42 per cent of GDP to about 49 per cent. The immediate source of the problem lies, therefore, in the huge demands of the Public Sector Borrowing Requirement. In the medium-term La Malfa plan (1981-1984) this was set at 10 per cent of GDP for 1982, but the figure achieved was 13 per cent, and the 1983 figure, despite a revised target, was forecast by the Treasury in February 1983 at nearly 17 per cent. But when the government presented a set of decree-laws to parliament attempting to control the expanding deficit in the same month, its original proposals were radically altered: proposed increases in prescription charges were reduced, new exemptions to charges were introduced, proposal reductions in sickness benefits were curtailed.

A further, separate problem is the habit parliament has developed of passing laws without adequate financial cover. In the period prior to the June 1983 elections, the President of the Republic finally refused to sign two bills passed in this apparently unconstitutional manner, and returned them to parliament. It is not clear whether he is prepared to do this with all such bills in future, nor indeed why he should have chosen those two measures out of many examples.

Italy

The problem is not only that the government is
unable to get its measures on welfare benefits and
costs passed intact by parliament or that parliament
insists on passing its own measures: the government
is also unable to control the current spending of
its own ministries and agencies, many of which, as
the 1983 Report of the Court of Accounts showed,
regularly overshoot their budgetary targets on the
current account, though they underspend by a lesser
amount on capital accounts. The gaping hole in the
state's account is accentuated by the effects of the
recession on tax revenue, but the consensus among
policy-makers appears to be that direct taxes are
already sufficiently high and cannot be increased
without risking a tax rebellion.
Finally, it should be noted that for some time
Italy has had relatively high interest-rates which,
at a time when inflation is actually declining,
albeit slowly, reflect not only the difficulty of
financing the public-sector deficit in the face of
increasing resistance of savers to government stock,
but also the need to maintain an effective differential against foreign rates to provide some cover
for balance-of-payments deficits. These two factors
make it difficult for the authorities to use interest-rates to control domestic liquidity.
However, granted that fiscal policy and other
instruments such as wage freezes are ineffective,
the authorities are forced to rely on the heavy hand
of monetary policy in its various forms to fulfil
major functions in macro-economic demand management,
mainly, in recent years, that of dampening domestic
demand. At the same time, the reliance on a high
level of public spending (among moderate realists)
for the purposes both of supporting the benefits-
transfer system and of inducing structural reform to
improve productivity (as in the La Malfa 1981-1984
plan) appears to be a strategy completely at varia-
nce with such a use. With the present structure of
Italian political economy, this contradiction is one
that governments have to face repeatedly and in
present political circumstances are unable to resolve.

Social Policies

There is a variety of ways in which positive
choices may occur in Italian politics - for instance
in parliament, or by independent agreement among the
parties, or by relatively independent decision of an
administrative organ such as the Bank of Italy.

Italy

Rarely does the Council of Ministers take the lead as a policy-determinant, and its role is often confined to ratification of proposals which originate from individual ministries, sometimes without even a minimal attempt at coordination. The Council of Ministers may sometimes appear to be the formal and final arbiter in major disputes over governmental budgets, but in these cases the ministers are clearly operating within a political brief from their party secretaries or national executives; the party secretaries, particularly of the PSI and DC, do not often take ministerial posts. The major impulse for specific policy choices, if these are radical or signify a change in direction of policy management, will usually be the result of a convergence of many different pressures, many of which come from alternative elites, international agencies, or foreign pressure. But there are also cases where radical reforms have been undertaken, or attempted, as a result mainly of domestic and intra-governmental developments. Two fields, health and education, are significant in their own right and for what they tell us about how policy is made in the Italian political system.

Health Policy

A law passed in 1978 was intended to be a progressive and radical attempt to introduce into Italy a National Health Service in deliberate emulation of and improvement on the British. Five years later, one of its main proponents, the PCI spokesman on health, admitted disappointment and Bettino Craxi announced that one of his new government's most urgent tasks would be a reform of the reform, with the particular aim of bringing the cost of the service under control.

Before one can udnerstand what went wrong, it is necessary to understand what problems the 1978 reform was aimed at and why it occurred. Just as the regional reform of 1968 was a belated response to a section of the Constitution, so the health reform was, to a considerable extent, forced on the then government by earlier legislation. The first key in the lock was a law of 1968, passed at the end of a parliament dominated by centre-left coalitions, which had attempted to bring financial equilibrium to the jungle of hospital care by establishing the responsibility of the *mutue*, the state-supported insurance companies, to meet hospital bills in full. It also established the right of all patients to

treatment in hospital, whether they were insured or
not, and it set up boards to oversee the efficient
management of the hospitals. One of the results of
the law was to make the hospital boards into centres
of wasteful clientelism, because of their independence from political or administrative control and
their responsibility for personnel, equipment and
buildings.

The financial burdens imposed on the *mutue*,
themselves no paragons of efficiency, were worsened
by a rapid icnrease in the costs of the service, in
particular of medicines, in the period after the
reform. In part this was the fault of the health-
care system itself, since the financing methods
encouraged a quantitative approach to medical care
on the part of general practitioners. But it was
also a result of the vulnerability of a system such
as Italy's, without proper control of costs or of
health-care policy, to the pressures of the multi-
national drug companies. There was similarly little
central direction on hospital admission and stay
policy. By 1976 the health system was functioning
only because of increasingly large annual subventions from the government. But it would be mislead-
ing to assume that the gross inefficiency of the
health-care provision, the damaging effect of the
1968 reform in opening up the system without
providing proper organisational support, and the
rapid increase in costs were sufficient conditions
for the 1978 reform.

What appears finally to have compelled the
government to agree to the reform in 1977 was a
series of short-term political factors. First, there
was the imminent deadline established by a 1974 law
for the dissolution of the hundreds of *mutue* and
their absorption into a drastically-simplified,
centrally-controlled insurance system; second, a
1975 law had also ordered the suppression of redundant para-state agencies, many of which nominally
had health or welfare fucntions; third, the regional reform of 1968 had ordered the transfer of health
provision to the regions, and this had already
begun, with a final deadline of 1979; and finally,
the government was under pressure from the PCI,
whose votes were essential for legislation, to
produce a sign of good faith quickly, particularly
since the austerity programme of the time was
accepted only reluctantly by PCI activists and trade
union members. Socialists and Communists believed
health-care reform to be particularly urgent not
because of the impossibility of sustaining the

financial burden, but because they feared that the existing system was in fact changing in a regressive manner under the pressures of the 1968 reform, and that inaction would result in a further surreptitious reversion of much of the system to privately-controlled health care, with an increasing tax burden on the lower-paid for the remaining state-run structures.

The principles of the new law were to establish a single, uniform level of care over the whole country; to give priority to prevention over cure; to deconcentrate control of institutions to the regions and include political participation at regional and local levels; and to maintain public spending on health within tolerable limits. The backbone of the reform was the establishment of 670 Local Health Units with a wide range of functions including health education, environmental, school and occupational health, maternity and infant care, as well as general medical practice and hospitals. The reform was therefore a kind to satisfy both the PCI and the regions; the DC resisted initially, not least because of the threat to its patronage exercised through the *mutue* and because of the fear of the Church that its provincial welfare agencies would lose independence, while the medical profession also objected to the potential interference of politicians and administrators. In the final vote the right-wing of the DC abstained, and the compactness of the PCI and PSI votes was sufficient to ensure a large majority.

In some areas, the health reform appears to have been at least a qualified success, but even there the same criticisms appear as elsewhere: that though the principles of the reform may be sound, its application has suffered from shortage and misdirection of funds and from gross political interference in administration.

Throughout its legislative career, proponents of the reform resisted giving firm estimates of its cost, some even arguing that it would represent a net saving since it involved thorough rationalisation of existing structures. This ambiguity over its real cost was a crucial factor in making the reform palatable to the 'extreme realists' among the DC ministers, and after the reform was passed they were able to turn the argument against the reformers when its transpired that more finance was needed in the short-term. In general what may be said to have happened is that the government has failed to make available adequate funds for the short-term restructuring (despite the ambitions of the Health Plan produced by the interministerial planning committee), while the immediate paymasters, that is the regions, have spent increasing amounts simply to meet the demand for health care, without being able seriously to restructure so as to allow the Local Health Units to function properly.

As a result, those who supported the reform claim that its main objective, the direction of health policy away from hospitals towards society, has not been seriously worked for, and the real financial burden of health provision has continued to rise. But it should also be said that the regions themselves have usually shown an unwillingness to implement or even to produce long-term plans, partly because of the political difficulty of inherited problems: overmanning, absenteeism, waste, duplication and corruption, as well as the physical degradation of many inner-city hospitals.

The other major problem is that in an effort to put health-care under democratic control at the local level, the reform established that the Units should be responsible to boards composed of locally-elected members. However, since election is by nominees of local govenrment councils, the Units have rapidly become an important part of the patronage network: together with the directorships of local transport companies, cooperatives and other bodies, their management posts are distributed on the basis of party strengths in the area. Sometimes this system produces honest, competent, hard-working administrators; sometimes it does not. Other forms of dysfunctional political interference may occur at an earlier stage of proceedings - the building of a hospital to provide employment in the area, or to bring funds and prestige, for instance. Italy does not appear lacking in hospital beds or health personnel, but this mismatch of resources to needs and the mismanagement of those resources have not been resolved by the reform; neither has the control of public spending been assisted.

Education Policy

In the field of education policy, the pressures of socio-economic change, together with immobilism and mismanagement, have brought the system to a permanent state of spectacular functional incoherence, similar to health policy in the mid-1970s, but in education the radical reform which was passed for better or worse in health care has not been forthcoming. It has been discussed since at least 1958, and there is widespread agreement on its urgency, but this is not sufficient to ensure coherent action. Instead, there has been a series of partial measures, particularly at the junior secondary level (ages 11-14) which have left certain fields (such as primary schools) untouched and which have created problems for higher education.

Great progress has undoubtedly been made in improving standards of literacy in Italy: in 1951, an estimated 60 per cent of the population was at best semi-literate, that is, had not achieved the primary school leaving certificate; by 1971, this number had decreased to 32 per cent, including 5 per cent estimated to be illiterate. This has been achieved by the relatively successful enforcement of compulsory schooling up to the age of 14; but the system relies on a very large teaching staff (about 720,000 in 1977), of whom a large proportion (over 40 per cent) is temporary, merely covering for the sickness and considerable absenteeism of their more fortunate, permanently-employed colleagues.

While spending on salaries is very high (over 90 per cent of total expenditure on education in 1977), buildings and equipment have suffered badly from shortage of funds, so that schools have to have recourse to double and sometimes triple shifts for classrooms, and are compelled to use inadequate rented accommodation. As in health-care, these serious problems of management of basic resources are compounded by a lack of effective central control of the budget. Though the Ministry of Education pays considerable attention to the curriculum and timetable of schools, its consideration of the financing of education is limited, in the formalist tradition of the Italian bureaucracy, to checking that actual spending corresponds to the proper categories and follows the correct legal procedures. The distribution of the funds into the categories is a matter for administration not politics, and it reflects both the inertia of a prestigious and long-established ministry and its vulnerability to vested interests - in particular, to the primary and secondary school teachers and to the 'professor-barons' of the universities.

Mismatch between the output of the educational system and the demands of the national economy is not unique to Italy, nor is it a recent development. Since the Risorgimento, the need to socialise important sectors of the population into support for the political system, the demands of the middle class for upward social mobility, and the capacity of the education system to produce large numbers of graduates and skilled secondary-school leavers, have come into conflict with the limited capacity of the economic system to absorb such people. Though the Casati Law (1859) emphasised the importance of using education to integrate the peasants and urban working class into the new state, the education system that developed was a relatively closed one, centrally controlled, limited in the social classes for which it catered, and with a strong emphasis on education in the humanities.

In the 1960s, two major but uncoordinated reforms destroyed this general cast of policy: in 1962 the traditional distinction in the junior secondary school between classical and technical schools was replaced by a fully comprehensive system for 11-14 year-olds, and then in 1969 the universities were compelled by law to accept all students who had passed the senior secondary school leaving certificate, irrespective of the faculty they wished to enter and of the particular discipline (technical or classical) in which the exams had been taken. One effect of these measures was to reinforce the tendency for higher education to become a parking-lot for well-qualified, unemployed school-leavers. Little effort has been made to provide facilities for the large increase in university students. Universities are unable to function properly because of inadequate accommodation; student-staff ratios are high because of inadequate staff numbers and, because of the frequent absenteeism of tenured staff, consistent staff-student contact is difficult if not impossible to achieve in most faculties; the examination system, based largely on a treadmill of frequent oral examinations for four years, followed by a virtually unsupervised thesis in the final year, is an ineffective means of testing the students' capacity.

In short, Italian education seems unable to respond either to demographic and socio-economic change or to the effects of the world recession. In 1962, there were about 225,000 students attending university full-time in Italy, with 24,000 graduating. By 1978, the number of first-year registrations alone was 227,000, the total number of students 750,000, and the number of graduates 77,000. Increases in student numbers were occurring throughout Western Europe in this period, but in Italy the increase was particularly large, was not matched by a corresponding rise in capital spending, and took

place in a completely haphazard and uncontrolled
manner, without serious practical consideration
either for the capacity of the university system to
educate such large intakes or for the capacity of
the economy to find work for its numerous graduates.
Unemployment among graduates is increasing: the
title of *dottore*, claimed by all graduates, was once
regarded as the key to secure employment and certainly for the lower middle and working classes as a
guarantee of upward social mobility; but the economy
is now unable to soak them up.

Unemployment is a problem particularly for
younger age-groups. In 1981, of 1.9 million unemployed, about 1.4 million were in the 14-29 age-group,
1.2 million under the age of 25, and 419,000 of them
graduates or holders of the upper-secondary school
leaving certificate. Not surprisingly, granted the
failure of government employment programmes and the
relative success of the unions in preserving job
security in the established industrial and services
sectors, the majority of these young unemployed have
either no experience of paid employment or very
little.

The field of education has not been without its
efforts at ameliorating an anachronistic and un-
wieldy system. A decree-law of 1973 on the status
and organisation of teaching staff which allowed
the Minister of Education, behind its apparently
innocuous title, to institute a completely new management system for schools, based on democratic representation of all groups involved including parents.
In 1979 there was a modification of the curriculum
which particularly affected junior secondary schools
and removed some of the residual classical elements
left by the 1962 reform. There have also been reforms of the status of university staff and of university curricula. But the problems left untouched by
such incremental efforts are large: the primary
school curriculum was last changed in 1955 and dates
predominantly from 1928, while the crucial upper-
secondary curriculum, though the subject of many
reform attempts, is substantially unaltered since
1923.

The question why it has proved possible to pass
a thorough reform of an anachronistic and dysfunctional system in health but not in education cannot
be quickly answered; the question becomes even more
opaque if one considers that education is one of the
traditional and essential functions of the liberal
state, and even in the post-war period has occupied
and continues to occupy a good deal of attentions of

politicians and administrators. The debate over education in Italy has a very lengthy if inconclusive history, and since the 'year of the student' in 1968 numerous proposals have come from all part of the political spectrum. Typically, the response of the government to the student protests of that year was acquiescent, tactical, and inconsistent: though the climate of discontent made the debate on education policy even more urgent, the resultant half-hearted measures in the 1970s satisfied none of the influential groups.

The high profile of the issue may be one of the most important reasons for failure, together with the ideologically-charged character of the problems. The major parties were in fundamental and highly public disagreement over the proper objectives of the education system, and the reforms democratising school management have accentuated the importance of this factor. While the extreme realists, particularly in the Liberal and Republican parties, insisted on the demands of the economy for a limited number of technologically-oriented graduates, the Catholic residues of the DC pushed it towards the whole and healthy development of the individual personality, recognising inequalities of ability. The Communists were more prone to argue for a materialistic approach involving both 'the logic of material production' and the wider framework of society within which the logic is held to operate.

Though a bill for upper-secondary school reform was produced by the Education Committee of the Chamber of Deputies in 1976, the slowness of its progress reflected the lack of consensus among the parties, and its committee origins indicated the unwillingness of successive governments to produce a coherent, acceptable and orceable policy. After a hiatus in the 1976-1979 legislature, caused largely by a complete breakdown of consensus between the PCI and DC over the issue, the Chamber passed a bill based on the 1976 text in 1982, which was still proceeding through the Senate in September 1983.

Finally, a major reason for the immobilism has been that the two most influential forces for change in other fields in recent years - the trade unions and the new regions - have been relatively inactive in the debate. The unions successfully fought for the introduction of workers' rights to further education in 1973 by direct negotiation with the employers, but have not otherwise given education a high priority. The new regions are responsible for further education and professional training, and are

Italy

supposed to administer the schools building programmes, but have little other scope for the exertion
of the kind of pressure which was possible in health
policy. Without the unavoidable pressure of their
own legislative deadlines for devolution, governments have not taken the lead in education. In this
field the commonplace style fo the DC has predomin-
ated.

Conclusion

This chapter has argued for a perspective on
policy-making that gives due weight to international
and domestic contraints which are specific to Italy
and that recognises the inertia of the Italian political system. The period under consideration has
been characterised by enormous changes, both nationally and internationally. Nationally, political
debate has been radicalised, there have been phases
of intense public involvement in political events,
and major new political actors have made themselves
felt. Internationally, Italy has suffered increasing
economic instability following the breakdown of the
international currency system in the early 1970s,
from the oil price rises in 1973 and 1979, and from
the world recession since then. But these factors,
many of which Italy shares with other Western
European countries, have not led to a uniformity or
a consistent pattern in response. On the contrary,
what emerges is the diversity of the policy reaction
in different fields and the different effects which
policy has had.
The debate has generally centred not on political stability, or economic growth, or even maintenance of achieved standards in specific fields, but
rather on the need to reach a stable and efficient
administrative arrangement. However, it has also
been possible to pass major reform measures not
directly related to efficiency in the state apparatus, and the energy crisis or the impact of the
recession. In part, this may be due to the lack of
electoral salience of the issue of administrative
efficiency, but a contributory factor is also the
increasing openness of the legislative process to
diverse influences. The various changes referred to
throughout this chapter have not fundamentally
altered the way politics proceeds in Italy, nor do
they appear to have stimulated major reform, though
they have certainly sharpened certain problems, have
made some prospective solutions unattainable, and
have modified the roles played by some of the actors,

151

in particular the DC.

The pattern that emerged in the late 1960s was one that allowed an ever-widening group of participants access to a diminishing and crumbling array of social and political goods, though that access was controlled with varying degrees of rigour by the dominant Christian Democratic Party. It has been argued by some that this capacity of the DC to tolerate all comers by changing all comers is at the centre of the problem of 'the available state' and is the major reason why there is no strong motivation on the party of any major political actor seriously to introduce reform of the state. There is undoubtedly accuracy in this argument, but important qualifications must be made to it. First, a sense of historical continuity must compel us to recognise that there have actually been serious attempts to change how things work (or do not work) in Italy - first by the Socialists in the centre-left period in the 1960s and secondly by the PCI during the 'historic compromise' parliament. A comment by Giorgio Ruffolo on the first of these periods is appropriate to both: *"The centre-left's reforming action in Italy shows the capacity to inflict deep rents on the conservative fabric by adopting important and daring innovations without, however, fitting the latter into a new and consistent texture of programmes, consensuses and institutions."*

The flexibility of the Christian Democrat system of power, based as it is on a variety of kinds of electoral representation and on a variety of ideological sources, is such that it has proved able in the long term to adapt itself to what were 'important and daring innovations' in ways that ensure its own survival. The second point to be made is that there may be a very wide gap between declared governmental intentions and the content of official acts, and then again between official acts and the real acts of the administration. The distinction between the *paese reale* and the *paese legale* is deservedly one of the banalities of Italian history. The distinctions to be made are more complex. In terms of policy output, the important conclusion is that public spending and public policy in general may depend more on the automatic mechanisms established (albeit involuntarily) by the law or governmental act than they do on declared governmental decisions. This is particularly the case in a country such as Italy which has a formalist legal tradition in which the function of public administration has generally not been the maintenance of

Italy

rational managerial control over public policy in
accordance with government intentions, but rather
the control of the formal legality of procedures.
 In conclusion, it may be said that the distinctive characteristics of the Italian response to
current international problems of political economy
have been in conformity with the established pattern
of policy-making, marked by an increasing diversity
of approach over time and in particular fields,
against the background of DC dominance of the state.
This dominance may now be weakened with the advent
of non-DC Prime Ministers, but they remain the largest single party, and the residue of their 35-year
control of the state is large.

Chapter Six

SWEDEN

Gunnel Gustafsson & Jeremy J. Richardson

In order to identify and analyse the central
characteristics of Swedish policy and politics it is
first necessary to describe briefly some cultural and
structural phenomena of vital importance in a comparative perspective. The main task of this chapter is,
however, to examine Swedish policy processes and some
of the key policy issues in Sweden today. The issues
considered in some detail are the 'hard-core' components of welfare politics, i.e. tax policy, labour-
market policy, and social security policy. Nevertheless, an unconventional demand dimension began to
emerge in the 1970s. It is difficult accurately to
label this activity, but it might be characterised as
a 'life-style' movement, emphasising non-material
values, protesting against, among other things,
excessive use of technology and large-scale societal
solutions, and believing that there was insufficient
concern with 'higher-order' needs like friendship and
solidarity. No doubt many Western democracies developed a concern with these post-material values during
the 1970s, but it is fair to claim that the debate
has been unusually intense in Sweden. In discussing
the hard-core issues, it is important to note that
these newer issues were an important part of policy
debates. Some of them, such as nuclear energy, were
very stressful - this one, indeed, was the subject of
a national referendum in 1980.

Cultural and Structural Patterns

In Swedish political culture there is not a wide
conceptual difference between the public and private
spheres of life. Sweden has never been characterised
by the British tradition of social-contact liberalism,
in which government is justified insofar as it is
necessary to protect the rights of individuals.

Instead, government exists as a means for the achievement of the public interest or of a good society. Government has become generally accepted as a 'helpful father' who is both willing and capable of taking care of life as a whole. Critical foreign observers have commented that Sweden appears to have a law for everything - even governing the colour which houses may be painted and whether you can smack your children!

The cultural integration of the individual and the state has a series of implications of great importance for the means by which collective measures should be undertaken. One example is the effort to formulate preventive policies. The Swedish approach to policy-making has tended to be anticipatory rather than reactive. It tries to avoid having defined problems suddenly thrust on the political agenda. Instead, Swedish policy-makers usually act in order to prevent something which is otherwise likely to need future correction. This is a long-term agenda-management strategy. It means that future perspectives are part of the policy-making routines. There is a tendency to produce forward plans, presenting ideas on how to formulate broad policy goals. These plans are produced in regular cooperation between politicians, bureaucrats, clients and affected interest groups. There is a concerted attempt to develop a consensus on what the future may bring and on which policy options should be attempted.

This preventive or anticipatory attitude, closely linked to the identity between the individual and the state, can be seen as a main ingredient in the radical/consultative policy style which has made Sweden a welfare-state model. The actual *manner* in which Sweden is governed today is quite different from that of a few decades ago. Cultural orientations towards compromise and consensus are, however, still dominant. In the current absence of a more fundamental agreement about future policy goals, the focus on consensus is often the explanation of incremental rather than radical changes in policies. Incongruences between patterns of conflict and structures aimed at producing compromise lead to diffusion of power, which in turn means more symbolic and pseudo-policies.

In the long run this produces a schizophrenic political culture. The official picture (and cultural value) is characterised by rationality, consensus, long-term orientation and common efforts to reach the general good. The other side of the coin is ready acceptance of unpredicted developments, consensus reached not by compromise but by postponing or other-

wise avoiding issues of conflict, and sophisticated
fights by specific interest groups in the name of the
public good.

Since 1975 Sweden has had a unicameral parliament of 349 members. Elections are held every third
year, with proportional representation of all political parties which get either 4 per cent or more of
the total vote or 12 per cent of the votes in a
district. The Swedish multi-party system is an
extremely strong, stable, and important part of the
governmental structure. Since the 1930s parliamentary
representation has been confined to the same five
parties: Conservative, Liberal, Centre, Social Democratic and Communist. Minor parties have, however,
tried to win seats. Most important of these are the
Christian Democratic Union (KDS), which for some 20
years has won 1-2 per cent of the votes, and the
Environmental Party which first ran in the 1982
election and achieved 1.65 per cent of the votes.
However, the 4 per cent hurdle makes it difficult for
small parties to achieve representation. When the
non-Socialist parties (Centre, Conservative and Liberals) gained a parliamentary majority in 1976, the
Socialists had been in power for more than 40 years.
During the 1970s Social Democrats did not, however,
have a clear majority. In the period 1970-1973 they
governed with the support of Communists; in 1973-1976
the two parties together held exactly half of the
parliamentary seats. After the 1976 and 1979 elections
the three non-Socialist parties formed coalition
governments which on both occasions collapsed before
next election. In 1978 the disagreement concerned the
nuclear question, after which the smallest of the three,
the Liberals, formed a minority government for a year.
During the period of 1979-81 the coalition split was
caused by disagreement over tax policy, with the
Centre and Liberals forming a minority government.
The Social Democrats were returned to office in 1982,
but have again to rely on support from the Communists.

In general, the five 'old' parties are slow to
adjust to social and ideological changes in the post-
industrial society. The electorate is starting to
divide in other ways than the traditional left-right
dimension. This is reflected in the nuclear question
and more generally in the non-materialist movement.
As a result, new cleavages are present within all the
traditional parties, making it much more difficult to
govern Sweden. The smaller parties play a more significant role at regional and local level where the
party system is also important. For example, the subnational party organisations are very important in the

recruitment of politicians. During the last elections,
turn-out at the municipal level was more than 90 per
cent, as high as at the national and regional levels.
    Sweden is usually considered to have a highly
centralised system of government. However, there have
always been strong local government units as well as
strong administrative boards such as the Education
Board, the Agriculture Board and the Social Welfare
Board, usually organised at national and regional
levels. According to the Swedish Constitution, parliament should formulate and coordinate national policies
to be implemented by the administrative boards,
regional and local authorities. There is a distinction
between ministries which have the major financial and
political responsibilities but are quite small, and
administrative boards which are large and numerous
(around 80) autonomous bodies responsible for prepar-
ing and implementing political decisions. Over the
last decade the trend has been towards decentralisation of the power: in many fields (for example,
education, housing, social welfare and environment)
there is now greater capacity to determine policy at
regional and/or local level. Together with a strong
centralised state there has, however, always been
strong decentralisation and segmentation. Today, the
roles of different agencies are much more blurred and
the political-administrative system is not only
extremely complicated but is strongly characterised
by diffusion of power.
    Recent local government reforms (amalgamation
into some 280 municipalities) have meant that local
government units have become even stronger. Two
points which contribute to the key role played by the
local authorities may be made. The first is the right
laid down in the Constitution entitling municipalities
and counties to levy flat-rate income tax (unlike
national income tax which is on a progressive scale),
with local rates decided by the councillors. Municipal
and county taxes have increased continuously over the
last decades, the mean rate rising from 12 per cent
of taxable income in 1955 to almost 30 per cent in
1981. The second is the degree to which citizens are
dependent on municipal services. Several important
tasks which in many other countries are performed by
national government are the responsibility of local
authorities in Sweden. Local authorities have a general power to conduct 'their own affairs', which
means they may undertake any 'appropriate' action
their councils deem to be in the interest of the
inhabitants of the area. A resident may appeal to a
higher authority, in the last resort the Supreme

Administrative Court. Thus the limits to what can be
decided in a Swedish municipality are not at all
clearly defined. If the higher authorities find that
the general power is being used unreasonably, the
decision in question is rescinded, but this does not
prevent other municipalities from doing the same
thing until successfully challenged by their own
residents. The views of the citizens thus often
determine the range of activity of their own municipality - rather than the activity being strictly
defined by national legislation. In addition, there
are fields of authority conferred by specific legislation, e.g. the Education, Building and Environment
Laws. Here duties are imposed on local government
and these functions are often financed by means of
state grants.

Welfare Policy and Politics

During the creation of the Swedish welfare state,
i.e. from the second world war until the mid-1960s,
there was rapid expansion in the economy. The Social
Democrats were in a powerful position and they built
their welfare society on three main corner-stones,
namely progressive income taxes, in the main financing
an advanced social security system, and full employment. Thus, labour-market policy, tax policy and
social-welfare policy formed an integrated system of
activities well suited to creating the Swedish welfare
model. All three components were politically controversial, but the long Social Democratic dominance in
parliament and government allowed sufficient time for
gradual implementation and absorption of social
democratic values in the whole of society.
During the 1960s, however, young people began to
doubt materialist values based on a high living standard. New demands were made on the political system.
These were quite diverse but generally emphasised
ecology, peace, equality (between developed and under-
developed countries, men and women, children and
grown-ups), etc. Public debate was particularly concerned with decentralisation in a broad sense. The
Centre Party version of the new movement meant opposition to nuclear power, arguments in favour of small-
scale solutions to society's problems, and less
emphasis on moving people to large towns. Spokesmen
of the new life-style, protesting against the concentration of power (in political, bureaucratic or
economic circles) and stressing self-determination
were, however, to be found among all party supporters.
During the 1970s the divergence between traditional-

ists and non-materialists became quite visible within all five main parties, not least in the Social Democratic camp.

While economic growth continued, the expansion of the public sector also continued and was not really a serious political issue. On the contrary, it was seen as a guarantee against unemployment, as well as an efficient means of extending the welfare state. Social Democratic governments continued to use the old policy solutions and to build on the long tradition of extensive consultation. However, the combination of the new post-materialist demands and the gradual increase in the size of the public sector (as well as other factors) produced incremental changes in the policy process itself. It began to be characterised by, for example, unconventional participation and overcrowding, making it difficult for governments to implement policy change. Some observers saw Sweden as developing a new version of the 'middle way'.

The non-Socialist parties became more vociferous in their criticism of what they saw as damage to the market economy. This criticism was levelled at the three policy areas which constituted the main elements of Social Democratic welfare policy. The social security system was considered 'too safe' and too costly. The consequence of this, they argued, was a lowered willingness to look for work and therefore increased costs for labour-market policy. The tax system was also seen as dysfunctional from this viewpoint. The individual's willingness to contribute both to the public good and to market efficiency was being hampered by very high marginal rates of personal taxation.

The election of a non-Socialist government in 1976 coincided with the international economic crisis, which Sweden was unable to avoid. One can speculate about the erosion of Social Democratic voter support during the 1970s. There is good evidence that the main explanation is the decrease in the number of manual workers, i.e. a post-industrial or service society replacing the industrial society. The actual economic situation was, of course, also relevant. The basis of the Swedish model rested upon the three corner-stones mentioned earlier, and on the accommodation of all affected interests. This presupposed economic growth. When it became difficult to satisfy all interests by expanding the welfare state, Social Democratic support began to decline. The connection between political and economic developments in Sweden seems perfectly reasonable, as election studies show that class-voting over time has become less important. According to this hypothesis, it is possible to view

Sweden

the increased support for Social Democrats in the
1982 election as a 'disappointment reaction' on the
part of voters. The economic situation did not
improve during the six years of non-Socialist governments. On the contrary, it worsened. Nor did Centre
Party supporters, sympathetic to the 'non-materialist'
movement, see many positive results of their party
being in government. Election surveys indicate that
Sweden is developing a modified two-party system. The
following percentages were reported in April 1983:
Communist-4.0, Social Democrats-45.0, Conservatives-
28.0, Christian Democrats-2.5, Centre-12.5, Liberals-
4.0, and Environmentalists-3.0. The two largest
parties are those which are least (publicly) in favour of non-material demands. It is, therefore, not
surprising that a new 'green' or Environmental Party
entered the election scene and gained some support,
(1.65 per cent) in the 1982 election.

Taxes

During the creation and expansion of the Swedish
welfare state, the debate about the tax system was
closely connected with both social policy and labour-
market policy. Social Democrats emphasised that social
reforms and full employment were possible through
increased taxes. The relationship between tax levels
and the possibility of further social reforms was
generally well recognised. The debate between Socialists and non-Socialists was, of course, over the
degree to which it was fair or reasonable to make high
income earners pay for advanced social reforms.
Until the end of the 1950s the Social Democrats
argued that indirect taxes were unfair because they
hit the poor more than others. Nevertheless, indirect
taxes began to be used on a larger scale, in order to
finance new social reforms. Indirect taxes are now
rather important. In 1981/82, taxes on income, capital
gains and profits accounted for only 18.6 per cent of
state revenue; social security contributions for 18.1
per cent; taxes on property, alcohol, tobacco, energy,
petrol, road traffic import duties for 18.7 per cent;
and revenue from central government activities for
10.4 per cent.
The early 1970s saw central government income
tax rates increase considerably for middle-income
taxpayers. In 1982 the direct tax burden (national
and local) of a married taxpayer with two children
(single income and allowing for child allowances) was
7 per cent of assessed income at £3,500; 22 per cent at
£6,000; 33 per cent at £8,600; and 56 per cent at

161

£17,000. At this time tax evasion and tax fiddling began to emerge as a problem. The decline in the economy made it clear, not only to the non-Socialists but also to Social Democrats, that the tax system had to be reformed.

It was felt that the new economic situation would erode the productivity of the economy, and would make it difficult to maintain the existing welfare standards, including full employment. In other words, the link between higher taxes and a better welfare system appeared to be broken. In particular, the marginal rate of income tax made individual earners less willing to contribute to collective goods and more inclined to focus on their private economy. The whole system began to be perceived as somewhat unfair. Citizens who paid high marginal taxes were not convinced that they were wealthier than those who gained from extensive welfare either through low marginal rates or high transfer payments and low income-based charges for services. As a result, during 1981, an agreement was reached between the two governing parties (Liberals and Centre) and the Social Democrats. The main thrust of this was lower marginal income tax rates, but also lower tax deductions for interest payments. The Conservatives left government as a result of this agreement.

The agreement did, however, leave many details to be settled. Negotiations were to take place in order to solve the question of financing the reforms. Its implementation has presented specific problems during the period of the new Social Democratic government. The two non-Socialist parties argue that the 16 per cent devaluation of the Swedish Kroner (decided by the Social Democrats during their first month in office in 1982) made the whole agreement meaningless. Social Democrats, however, argue that they will implement the tax reform as planned but have also pointed out that they have never guaranteed the 1981 material living standard. It still remains to be seen how the tax reform will be financed. During the spring of 1983 the government formulated a bill in which it is suggested that this should be done through higher taxes on certain goods, especially energy. The Liberals argued that because of the devaluation the tax reform was by then a joke. The Centre Party leader welcomed the fact that the Social Democrats avoided new taxes on firms. Conservatives, of course, stick to their original position and press for reduced public expenditure, especially on social welfare.

Sweden

Social Policy

Social policy includes activities which to a
great extent are not just regulative but distributive
and redistributive. At the service delivery level
there is, as a consequence, a mixture of political
and market allocations. The political system is as
important as a distributor of goods and services as
the market. Until the mid-1960s the non-Socialists
constantly criticized the Social Democrats for
carrying out social reforms which were too advanced
and too rapid. However, there were in fact relatively
few occasions when there were political conflicts
over specific aspects of the social policy programme
(one example was the supplementary pensions scheme
at the end of the 1950s).

By the early 1970s some contradictions in social
policy had began to emerge. Firstly, there was an
alarming rise in costs, especially for health and
social insurance. The increased costs for health
care are reflected in the very high increase in local
taxes, whereas the cost of the social insurance
system was divided between the state and employers
(with the latter responsible for 85 per cent of
sickness benefits and a third of pensions). Secondly,
social policy was fragmented in terms of responsibility for the definition of goals as well as for
actual implementation. Some parts of social policy
are processed by the state (in several ministries)
and others by counties and/or municipalities. It has
become very difficult for clients, politicians,
bureaucrats and others to see a clear picture of the
entire system of social policy.

As in many policy areas, organisational problems
are very significant. There is fragmentation and
sectorisation of the bureaucracy to such an extent
that it becomes almost impossible to identify where
responsibility with regard to a specific case is
located. Public undertakings have increased incre-
mentally as one reform, often aimed at solving a
specific problem, has been added to another. It has
often been necessary to create a new body or add a
new function to some already existing organisation
handling other matters. This has usually been done
without considering the entire public decision-
making or implementation system. The extension of the
public sector has, therefore, created coordination
problems. Attempts to solve these have been through
various types of decentralisation. One important
example is the new framework law on social assistance,
which makes municipal authorities responsible for

coordination to ensure that those in need get the best possible help.

During the first year of the second non-Socialist government, at the turn of the present decade, there was for the first time an alarming increase in the national budget deficit. (In 1975/76 the deficit was Skr. 3.7 billion, whereas comparative figures for 1978/79 and 1980/81 were 36.7 and 59.7 billion respectively.) This problem provoked a serious debate on the need for a less ambitious social policy, including cuts in subsidies and benefits. Within the government the three parties agreed that the national economy required cost reductions in the social welfare system. The view was, of course, consistent with ideology concerning the need to limit he public sector and the need to make the 'market' work again.

In 1982 the non-Socialist government proposed a three-day waiting period for sickness benefits. This suggestion, however, met very strong opposition, not only from Social Democrats and trade unions, but also from supporters (and even MPs) of the governing parties, particularly the Centre Party. The main problem was that such a reform could be seen as state intervention in the free negotiations of wages and related benefits between employers and employees. In fact, white-collar workers had already negotiated an agreement removing them from the proposal. If their agreement was upheld, then the manual workers were the only ones likely to be affected by the new policy. This difference of treatment caused great bitterness, particularly from the trade union confederation, and the public debate took place in an emotional atmosphere. Despite the criticism, the government managed to obtain a parliamentary majority for its bill. The Social Demcorats, however, promised (in the election campaign) to abolish the 'waiting days'. They also said they would guarantee the pension system and the state subsidies to municipal day-care activities. The Social Democratic victory of 1982 thus meant (not just in theory but in practice) that the social security system has remained relatively intact despite the economic problems. Instead of cutting social policy expenditure, the emphasis is now on an economic policy which, through devaluation of the Swedish Kroner, tax policy, etc., tries to revitalise the economy.

By international standards, Sweden's citizens remain well cared for by the state. In practice, there is a very detailed social-welfare programme, seemingly designed to cover almost any eventuality. Typical of the degree of fine tuning is the parental

insurance scheme under which, after the birth of a
child, one of the parents is entitled to six months
off work usually without wage reduction. In addition,
each of the parents is entitled to another three
months leave of absence from work, on a percentage of
gross income. Medical care, however, is not completely free. For example, there is a charge of just under
£3 for visits to a doctor and just under £4 for prescriptions. A patient may have to pay up to 60 per
cent of the costs of dental care, providing that
these do not exceed aproximately £200.

One sector of social policy, namely health care,
is, however, undergoing rapid changes. Again, there
are efforts to decentralise the system; and there is
also increased emphasis on preventive medicine. In
practice, this means that health care is increasingly
a county responsibility, both in terms of financing
and in terms of balancing the different (rather vague)
national goals. More emphasis on accessibility to
doctors and preventive care is meant to reduce hospital expenditure: the underlying idea (or hope) is
that less hospital care will be needed as a consequence of the new system. One interesting aspect of
the new policy is that there is now a debate about
home care instead of hospital care. A type of social
insurance, giving citizens the right of leave from
work to care for old and sick relatives, will probably be introduced. Such a system already exists as
regards child care, where parents have a right to
stay at home (with full pay) when their children are
ill.

In both the cases of child care and care for the
elderly, such 'deinstitutionalisation' of social
security is debated in connection with the 'non-
materialist' movement. Some look upon this trend as
a very good development in its own right. They consider it as a step towards higher-order values: an
increase in the quality of life. It might, however,
also be seen as a rather clever strategy, from the
viewpoint of traditional society, having implications
for sex equality. There is a distinct possibility
that this, and similar types of deinstitutionalisation or home care, will have quite unintended consequences. It might, thus, mean that the welfare system
is 'saved' by very traditional solutions, i.e. women
staying at home, reduced payments to young people who
are not able to find proper work on the market, and
reduced professional treatment of sick people and
other disadvantaged groups. The consequence of this
trend might possibl be a significant change in the
social welfare system, carried out by means of a very

sophisticated strategy. In many cases the divergence between the official purpose of such reforms and the actual results might sufficiently fudge the issue as to make it possible to build an initial consensus around what turn out to be radical policy change.

Labour-Market Policy

Labour-market policy has become one of the best known elements in the comprehensive welfare system in Sweden. The active intervention, by government, in the labour market is often cited as a model (as is West German training policy) for others to follow. Full employment is declared to be a central object- ive of economic policy, alongside economic growth, stable money, balance of payments and an equitable distribution of income. There has been a very con- siderable increase in public expenditure in support of the active labour-market policy. For example, during the 1970s, it consumed over 6 per cent of public expenditure, some two or three times higher than in the early 1960s.

The philosophy was that the maintenance of full employment by a non-selective financial and economic policy ran the risk of high inflation, and that selective, specific measures were therefore likely to be more effective. An essential element in the policy has been to attempt to match the labour force to job opportunities by a highly developed training and retraining system and by introducing incentives for the mobility of labour. These are the so-called 'matching measures'.

Social, as well as economic, goals have become rather important - for example, in the 1970s more policies were introduced to assist groups such as women and the handicapped who were less able to enter employment. In Swedish terms the shift was from 'full employment' as a goal to 'work for everyone' as a goal - a considerable heightening of ambition for the policy area as a whole. The political commitment to full employment has been extremely strong and Sweden still boasts one of the lowest unemployment rates in the Western world: in December 1983 it was a mere 3.5 per cent.

In a recent study Henning has argued that there has been a very significant shift, in response to the economic crisis, in the nature of Swedish labour- market policy.(1) In the latter half of the 1970s it was found that the traditional and well-tried labour- market measures were by themselves incapable of maintaining the very high levels of employment to

which Swedes had grown accustomed. In essence,
successive governments have found it necessary to
increase direct aids to companies in an effort to
encourage the retention of labour and to avoid
redundancies. Such measures have included stockpiling
subsidies, advance industrial orders and grants for
in-plant training of employees threatened with
redundancy. This long-term trend towards industrial
aids or subsidies is, however, he suggests, being
grafted onto the existing labour-market policies,
which remain a 'sacred cow' within the Swedish welfare sytem. There was a twentyfold increase in
allowances (i.e. subsidies, risk capital and loans)
to companies during the 1970s. Indeed, one study
suggests that subsidies to private firms were ninety
times greater in 1976 than in 1960. This very strong
trend was not affected by the arrival of bourgeois
governments, which have in practice exhibited a
progressive accommodation to the policies of the
trade-union and Social Democratic movements. Reality,
however, began to impinge in 1980-81 with emphasis
on the level of unemployment relative to other
Western nations and with a reduction in unemployment
benefits.
   The above developments suggest that there has
been a fundamental shift in the goal of labour-market
policy from the market as a whole to local markets.
The primary concern is now both the security of
employment and the location of the employment. This
has greatly reduced the flexibility of the labour
force, hitherto regarded as one of the great successes of the active labour-market policy. It also
appears that non-material values have begun to impinge on this well-established policy area. As
Henning's study says: *"Work has come to be expected
as a social right. The work has value in itself, and
as the material welfare has increased, the social
significance of the work has come, increasingly, to
be emphasised more strongly."*
   The main difficulty in the 1980s appears to rest
on two developments. Firstly, there is evidence that
attitudes may have changed - with less emphasis on
the need for economic growth and greater efficiency,
and more emphasis on staying in one's own community,
etc. Secondly, the well developed policy has quite
simply been unable to cope with the effects of world-
wide recession. As a result, short-term solutions
have had to be adopted of the kind quite familiar in
the UK and elsewhere, where public money, in ever
increasing amounts, is used to protect existing jobs.
This is not to suggest that the much vaunted labour-

Sweden

market policy has been a failure in the 1970s and
1980s. Just as, despite the stresses and strains, the
taxation system has not collapsed and the social
welfare system has remained intact, so this policy
has played a vital role in maintaining a very low
level of unemployment. One estimate is that for the
period 1974/80 labour-market measures diminished the
annual average level of unemployment by 3.2 per cent
in Sweden, compared with a reduction of 1.3 per cent
in West Germany.(2) It will, however, be extremely
difficult for even the Social Democrats to maintain
the effectiveness of labour-market policy (both in
the traditional sense and in the newer sense described above) in the face of a severe world recession. Social democratic values are, of course, deeply
entrenched and there is, as a result, a reluctance
to abandon previously cherished goals. Nevertheless,
economic forces are also difficult to resist, and
Sweden will gradually come to accept higher levels of
unemployment. As in all Western democracies, the
world recession has hit particular groups of workers
especially hard. For example, youth unemployment had
increased from a little over 4 per cent in 1970 to
nearly 8 per cent by the end of the decade. Total
'visible' unemployment was 3.5 per cent in the first
quarter of 1983, with 150,000 people unemployed.
However, over 180,000 people were on various labour-
market schemes and would have added an additional 4.2
per cent to the figures had they not existed.

Conclusion

In the opening of this chapter, it was suggested
that policy debate in Sweden has remained firmly
anchored in the core issues of welfare policies -
namely tax policy, labour-market policy and social-
security policy. It was also noted that Sweden had
seen the development of a 'post-materialist' movement
which had begun to challenge traditional values.
Though the movement can often be rather vague -
emphasising such values as 'togetherness' - it can
also generate quite specific policy issues - such as
the question of whether Sweden should continue its
nuclear energy programme.

Other new issues are more readily absorbed into
the traditional debate concerned the desirability and/
or practicability of extending the boundaries of
social democracy. For example, the increasingly bitter
discussion about the proposed Wage-Earner Funds,
though proving quite impossible to handle by means of
consensus-building procedures, is in a real sense a

further extension of participation rights for workers.
Thus, the management boards of the wage-earner funds
will consist of nine members, at least five of whom
will represent employees. Five regional funds, to be
set up in 1984, will invest in Swedish companies and
will, in essence, be providers of venture capital.
Their income will be derived from a levy on company
profits and from part of employers' national pension
contributions.

The wage-earner funds debate is a good illustration of the increased difficulty of achieving consensus on policy change in Sweden. In October 1983
Stockholm saw one of the largest demonstrations ever
known in the capital when 75,000 marched in protest
against the proposal, introduced by the Social Democrats earlier that year. The marchers included managing directors of Sweden's leading companies, as well
as owners of small and medium-sized firms. The
scheme's opponents see it as leading to the 'social-
isation' of industry and as fundamentally changing
the market economy. (The Federation of Swedish Industries estimates that within ten years the funds could
control 15-20 per cent of the total ordinary stock
quoted on the Stockholm Stock Exchange.) More significantly, the three non-Socialist parties have pledged
that they will revoke the scheme if they come to
office after the 1985 election. Such an example of
adversary politics, when the opposition sets about
undoing the legislation of its predecessor, is more
akin to Britain than to the Swedish consensual style.

There are other signs that the Swedish policy
process may have undergone rather fundamental changes
over the last decade. For example, there will be no
coordinated central negotiations for blue-collar-
worker wage contracts in 1984 between the Swedish
Employers' Confederation and the Trade Union Confederation. The tradition of effective central bargaining
had lasted for thirty years and had been internationally recognised as a great strength of the system of
wage negotiations in Sweden.

The economic situation is, of course, an
extremely important factor in structuring policy
debates. Sweden has not been able to side-step the
world recession, though this was its initial strategy. The deficit on Sweden's current account was
Skr.2,700m for the first quarter of 1983, with
foreign debts of Skr.138,000m. The harsh reality has,
therefore, encouraged a questioning of the degree to
which Sweden should maintain existing levels of
commitment to the extremely high standards of its
welfare state. The OECD forecast in 1981 that if there

were not fundamental changes in policy, Sweden's foreign debt would be so great as to threaten its credit-worthiness. There were some signs that the non-Socialist government of the time was beginning to re-orientate policies in an attempt to manage the economic problem. Like most Western democracies, Sweden faces the question of how to fund the ever-increasing costs of its welfare system. The projections are alarming. For example, a recent study by the Swedish Governmental Secretariat for Future Studies estimated that by the year 2,000 the system of public care will consume 20 per cent of GNP, compared with 14 per cent in 1983. This would cause a rise in taxation from 52 per cent of GNP to 61 per cent.

The secretariat sees the solution to this problem in the development of a system of sharing job opportunities. It suggests that: *"Shorter working hours, combined with community service and/or voluntary work in the care sector, can thus provide a solution both to the future unemployment crisis and the future care crisis. If we reduce working hours, nobody will need to accept a meaningless job for the sake of employment. If we share the work of care, everybody will be able to obtain all the care they need."* The report is, of course, merely a contribution to the debate and its implementation would present considerable practicable difficulties. It is quoted here, not because of its value as a policy document, because it neatly illustrates the possibility of a way forward for Sweden - namely the linking of solutions to the economic problems caused by the cost of its welfare state with solutions to the quest for non-materialist values in society. Sweden has had an enviable record of policy innovation in creating a welfare state. Circumstances may yet force it to be as innovative on the way down the slope as it was on the way up it.

Footnotes

1. Roger Henning, "Industrial Policy or Employment Policy? - Sweden's Response to Unemployment", in J.J. Richardson & Roger Henning (eds.), *Unemployment: Policy Responses of Western Democracies*, Sage, forthcoming.

2. See D. Weber, "Combatting and Acquiescing in Unemployment? Crisis Managmeent in Sweden and West Germany", *West European Politics*, Vol.6, January 1983, No.1, p.34.

Chapter Seven

OVERVIEW

Peter Walters

The picture of policies and politics in the five
countries covered in the earlier chapters of this
volume is of Europe in hard times. The impact of the
worst economic recession since the thirties is pervasive. In the Europe of the early eighties there are
no unambiguous success stories. Within each country
at least one economic indicator gives cause for concern. Even Germany, whose economy weathered the
seventies relatively successfully, is now faced with
a disturbingly high level of unemployment. Sweden may
have kept its unemployment well below the European
average but only at high cost to its exchequer; and
it remains afflicted by a higher than average inflation rate. Relative success, moreover, is small comfort to governments having to cope with the domestic
consequences of low growth. Shared pain is still pain,
and pain thresholds differ. The relativities which
trouble governments in traditionally successful economies concern present economic performance compared
to past experience rather than to other countries'
current plight. Thus, in Germany, permanently anxious
about the viability of its democratic institutions,
fears about the politicl consequences of a fall from
its post-war economic grace are particularly acute.
A similar theme is touched on with regard to the
implications of France's economic frustrations for
the still evolving and untested institutions of the
Fifth Republic. On the other hand, the apparent tolerance of the British polity to the consequences of
economic stagnation may be explained, partly at
least, in terms of Britain's long familiarity with
economic failure.
    The earlier chapters of this book reflect their
authors' sensitivity to the particularity of the
impact of straightened economic circumstances. They
interpret the general experiences of recession and

low growth in terms of its local significance. The
function of this concluding chapter is to put developments into comparative perspective. By drawing out
the common elements in the problems in the several
countries, it brings into focus such differences as
do exist in policy responses.

It is clear that in each country a similar
policy climate prevails. It can be illustrated by the
fate of Madame Nicole Questiaux, the French Minister
of Health and Social Security. *"I am Minister for
reforms not accounts"*, said Mme. Questiaux, brushing
off journalists' questions about her attitude to the
mounting deficits in the French social security
system. Her subsequent departure from office indicated the shift of priorities of the French government but was also symbolic of the constraint now
affecting policy-makers throughout Europe. Few ministers in any government can, in the present climate,
disregard 'accounts'. The precept of the Gladstonian
Treasury - finance comes before policy - is once
again dominant. This is manifest in the five countries' budgetary proposals for 1984, all variants on
a similar theme. In each the finance minister identified the need to control his budget deficit as the
central aim of fiscal policy. In each, the fiscal
discipline being stipulated is broadly similar, with
greater emphasis given to curbing public expenditure
than to increasing revenue. Whatever the political
complexion of the government in office, fiscal con-
servatism is everywhere justified as a fundamental
precondition of lasting economic recovery.

The preoccupation with reducing deficits reflects the significant retreat from Keynes which has
occurred since the mid-seventies. Governments do not
want, or feel unable, to revive their depressed
economies by stimulating domestic levels of demand
through an expansive fiscal policy. The precise
reasons vary, but there is a common thread in the
fear that fiscal expansion would prove inflationary
and damage trade balances. The short-lived French
'dash for growth' in 1981/2 is the exception which
in the eyes of more cautious governments has proven
the rule. Current European orthodoxy perceived fiscal
discipline as a necessary ingredient in controlling
inflation and fear of inflation has become a constraint which dominates economic policy-makers.

Most finance ministers would justify their
reluctance to expand their deficits on the grounds
that they are already too high: a serious impediment
to economic recovery because they necessitate high

interest rates which hold back private-sector investment. The most potent force uniting European finance
ministers is their common resentment of the huge
American budget deficit, the financing of which has
pushed up interest rates in Europe as well as
America. Given this interest rate constraint, the
general concern to contain deficits is understandable.
The automatic effects of recession is to widen deficits, pushing up expenditure on unemployment benefits
and cutting down tax revenue: a cyclical deficit is
a normal feature of an economy in recession. A remarkable feature of the early eighties, however, is
that certain countries, such as Britain and Germany,
have pursued fiscal rectitude to such a degree that
their budgets are in surplus when the cyclical deficit factor is taken into account. It is easier to
understand why Italy and Sweden should be attempting
to trim 'structural' deficits, which have reached the
level of 16 and 12 per cent of GDP respectively. In
fact, the concern which underlies the trimming exer-
cises in all countries stems from the common upward
trend of public expenditure over the past decade; a
trend which recession has accelerated rather that
caused. All governments, whether of Left or Right,
are troubled by the long term problem of financing
public expenditure in a period of low growth prospects without resorting to unsustainable deficits.
One alternative - increasing taxation - is a course
which governments are for various reasons reluctant
to take. The result, therefore, is a general attempt
to curb public spending; notably in the sector which
has in recent years grown most vigorously - social
welfare. Thus the significance of current fiscal
strategies extends beyond trends on economic management into the spheres of taxation policy and social
policy.

Economic Policy

Each country has followed a different road to
1984, but on their different roads they have encountered similar vicissitudes. A brief review of the
experiences and policy response of the five countries
since the mid-seventies shows how they have arrived
at apparently converging paths in the mid-eighties.
Examination of the wider national contexts of the
1984 fiscal measures, however, reveals significant
variations in ultimate objectives.
The background to the recent convergence
pattern is the succession of profound international
recessions which occurred in the seventies. The

173

first, in 1974/5, was exacerbated by the increase in
oil prices in 1973 which threw all the major econom-
ieces into balance of payments deficit and was both
inflationary and deflationary in its implications:
inflationary in the sense of adding substantially to
costs and deflationary in the extent to which it took
out domestic demand. Policy responses reflected the
priority given to the various facets of the problem.
The governments of the United Kingdom, Italy and
Sweden were amongst those which put fear of the
deflationary impact of OPEC actions first. Apprehens-
ive of the effect of a world recession on employment
levels, they were prepared to ignore the effect of
the price rise on the balance of payments and inflation, and they followed policies which kept home
demand buoyant. Germany's response, on the other
hand, reflected its overriding concern with curbing
inflation. It maintained tight monetary and fiscal
policies to limit inflation and prevent the balance
of payments going too deep into deficit.

The recovery from the 1974/5 slump was relatively weak. Growth resumed in most countries but not
at pre-1974 rates. Recovery was hindered by developments in some of those countries which had tried to
be expansionist in their response to the oil shock:
the United Kingdom, Italy, France and Sweden all
suffered accelerating inflation rates. Loss of competitiveness and weak international demand contributed to the continued deterioration of their balance
of payments. In 1975/6 all adopted restrictive
policies. In the second half of the seventies governments in those economies most vulnerable to balance
of payments constraint were beginning to argue that
renewed growth could not be internally generated
without aggravating their chronic inflation. They
urged the stronger economies such as Germany to adopt
expansionary policies which would 'pull' the weaker
ones out of stagnation by increased demand for their
exports. After the 1978 Bonn meeting of leaders of
the major Western economies, it appeared that this
strategy was being adopted by Germany and Japan. Its
potential success, however, was not put to the test.
The series of substantial oil price increases by the
OPEC countries in 1979 cut recovery short. Germany
responded with restrictive policies as in 1973. This
time most other countries fell into line, tightening
monetary and fiscal policy, intent on holding
inflation, safeguarding their balance of payments
and allowing unemployment to increase. This second
recession has been the deepest for the European
economies since the thirties. It has also been the

most prolonged with, in 1983, even the most optimistic politicians still concerned about the strength of
the recovery they claim to detect.

As well as being thrown off balance by the
difficulties posed by the severe trade cycles of the
decade, governments have become convinced that their
economies are afflicted by structural problems and
that these render ineffective, or are aggravated by,
counter-cyclical action. The OPEC price increases,
for example, brought home to European economies their
dependence on external sources of energy. This vulnerability is a constraint on the renewal of high levels
of growth because of fears that an increase demand
for energy will bring about further inflation and
balance of payments deficits. There is, moreover, a
growing belief that inflation and unemployment are
not just cyclical problems but the consequence of
structural features in European economies. These
include growing labour power, which has brought about
an endemic cost-inflation and has made labour markets
more rigid. The growth of public expenditure is seen
as starving the productive private sector. Moreover,
the Western economies, individually and collectively,
have become aware during the seventies of the obsol-
escence of extensive sectors of their industrial base
in the face of industrialisation in the developing
countries and the phenomenal success of the Japanese
economy. The need for industrial restructuring is
recognised in the strongest of the European economies. In all countries these problems have posed
governments a similar dilemma: whether to deal with
them by more interventionist policies, or whether to
rely on market forces to bring about the necessary
adjustment.

Britain

1976 marks a significant step in British economic policy because the new emphasis on curbing
inflation coincided with rising unemployment. It was
prompted by a combination of factors. The government's voluntary incomes policy - the 'Social
Contract' - was failing to bring down high wage
settlements and this contributed to inflation - 24
per cent in 1975. The fact that inflation was substantially higher than in other countries seriously
affected Britain's competitiveness in contracting
international markets. Output was declining, with
unemployment rising sharply in 1975 to the million
mark. With the balance of payments in substantial
deficit, there was little scope for expanding dom-

estic demand for fear of increasing imports. But it was the persistence of inflation at unprecedented and manifestly destabilising levels in the face of equally unprecedented unemployment which brought home to the government the need to abandon traditional demand-management techniques. The new 'realism' received its classic statement in Prime Minister Callaghan's response to the criticisms of the government's tolerance of mounting unemployment at the Labour Party Conference in 1976. *"We used to think you could spend your way out of a recession and increase employment by cutting taxes and boosting government spending. I tell you in all candour that the option no longer exists, and that in so far as it ever did exist, it only worked by injecting a bigger dose of inflation into the economy, followed by a higher level of unemployment as the next step."*

In 1976 the government encountered further limits on its options when it tried to bring about a fall in the sterling foreign-exchange rates in an effort to improve Britain's competitiveness. The rate fell uncontrollably because of nervousness in the foreign exchange markets about the state of the British economy. To break the fall, the government had to turn to the IMF for a loan, the terms of which locked it even more firmly into a deflationary stance. After the 1975/6 climacteric greater emphasis than in earlier deflationary phases was given to control of the money supply, which had come to be seen as a major factor in fighting inflation. The key was seen as the reduction of the budget deficit - the Public Sector Borrowing Requirement (PSBR). This tighter control of public sector spending involved cash limits and cuts in planned expenditure. By accepting the need to gear stabilisation policy to reductions in the deficit the Labour govenrment was acknowledging a major constraint on the role of the public sector in the economy: in effect, that its share of resources had to be reduced. This foreshadowed the Conservative government's policies. But despite the new emphasis on monetary control, Labour was traditional in its use of incomes policy to try to restrain wage inflation. In this it remained committed to a consensual approach to the problems of the labour market. When the Social Contract proved too permissive, the government negotiated with the unions in 1975 for tighter limits to wage increases, which prevailed with some success for three years. Like most earlier incomes policies, however, it proved unsustainable. Wage inflation accelerated in the 1978/9 pay round and the government's rather

ineffectual attempts to restrain public-sector wages
resulted in politically damaging strikes.

In other ways, the Labour government was radical.
Public ownership was significantly extended during its
period in office. The declining ship-building industry
was taken over, as were large sections of the areo-
space industry, on the grounds that they depended on
government investment and required major restructuring. The British National Oil Corporation established
a potent public presence in the new North Sea oil
industry. A National Enterprise Board provided investment capital for new industrial development. A role
for the state in stimulating investment in the private
sector was envisaged through Planning Agreements between companies and government. This general extension
of public power must be seen alongside the move to
restrict public consumption which was implicit in
fiscal policy. In a sense, the apparent contrast represented a cautious, pragmatic response to the problems of the economy: pushing forward the boundaries
of the public sector yet conscious of the need to
divert resources from consumption to production.

If the Labour Party, traumatised by having to
abandon its full employment commitment and to cut
public expenditure, was tentative in articulating a
new definition of the role of the public sector in
the economy, the Conservative government since 1979
has been less inhibited. Conservative policies have
been pursued with the vigour and zeal of a government
which considers its approach intellectually superior
not only to that of its political opponents but also
to that of previous post-war Conservative governments.
The essence of Conservative economic policy is a re-
definition of the responsibilities of government,
seen as basically to preserve the value of the currency and remove market distortions, thus providing
the essential conditions for an entrepreneurial
private sector to deliver employment and growth.
Control of inflation is not just a priority: it is
the only appropriate policy for government to pursue;
the only one for which government is directly responsible through its control of the money supply. Reduction of public expenditure is a necessary corollary.
Deficit financing is inflationary; taxation a disincentive to enterprise. Hence the key element in
Conservative policy over the past years has been the
more rigorous application of its Labour predecessor's
targets for monetary growth and control of government
borrowing. In practice, with money supply proving less
amenable to control than the government anticipated,
greater relevance has been placed on curbing expendi-

ture. This task has been made harder by the depths of
the recession in British which has cut revenue and
increased expenditure on social security. The position
has been eased by tax revenues from North Sea oil, but
the public sector's claim on GDP has increased,
serving to intensify the government's search for ways
of cutting existing programmes.

The government's strategic aim of reducing the
public sector's share of resources also reflects its
critical approach to the provision of public goods.
Thus, alongside the drive for cost-effectiveness
stimulated by the constraint of tight cash limits, the
thrust of its approach is to question whether a given
programme might not be adequately served by market
provision. The drive for privatisation of state economic enterprises is stimulated partly by fiscal
expediency (the need to cut public borrowing), partly
by a belief that private-sector management, because
it is more open to market forces, operates more
effectively than public management.

The Conservative government's aim of limiting
its role in the conduct of economic policy is evident
in its approach to the structural problems facing the
economy. In confronting wage inflation, it has explicitly eschewed a formal incomes policy, preferring
to rely on cash limits in the public sector and market
forces in the private sector to restrain wage levels.
At the same time it has sought to alter the balance
of power in the labour market through legislation
aimed at circumscribing the use of strikes. Encouragement of market forces is also evident in its policies
towards industrial restructuring. It has promoted
rationaliation with extensive closures and manpower
shedding in the nationalised industries. Only in its
encouragement of 'small businesses' has it developed
a new interventionist strategy in the private sector,
but here, too, the emphasis is on creating the
conditions for private enterprise rather than overriding the market. The government regards its economic responsibilities as restricted essentially to
the reduction of inflation and the stimulation of
market forces. The result is that in 1983 Britain has
one of the highest unemployment levels in Europe and
one of the lowest inflation rates. The renewal of
growth, argues the government, is a matter for management and workers to achieve.

France

French economic policy since the mid-seventies,
as that of Britain, has been notable for its discon-

tinuities. There have been two radical experiments:
the neopliberal restrictive approach to economic
management of the Barre government after 1974, which
questioned some of the traditional precepts concerning the the *dirigiste* role of the state, and the
expansionist policies of the Socialist government
after 1981 which extended the role of the public
sector. The Socialist's abandonment of expansionist
demand management in the course of 1982/3 represents
yet another change of direction.

The Barre policies in 1976 were a response to
the pressure of mounting inflation, a deteriorating
balance of payments, and fears of pressure on the
franc. Like the British Labour government at the
same time, the French government decided that the
problem of inflation had to take priority over unemployment. The result was a combination of tight
monetary and fiscal politics, with extensive controls
over wages. But the novelty of the Barre approach was
its assertion of the need for the private sector to
be freed from the heavy and sometimes misdirected
hand of state intervention. Future economic growth,
it argued, depended on the export potential of the
private sector: the state should try to reduce the
costs burden to industry and stimulate profits. Thus
price controls were lifted, wage controls introduced,
and non-investment public expenditure controlled.
This neo-liberalism, however, did not preclude the
continuation of the *dirigiste* tradition in selected
areas. Thus the government continued to promote,
through investment aid and sales promotion those
industries it considered had high growth potential.
Furthermore, it pushed forward with the nuclear
energy programme - the most ambitious in Europe.

The failure of the Barre experiment to stimulate investment and the resulting rise in unemployment
was a factor in the Socialists electoral success of
1981. The Mitterrand government expressed impatience
with the passive response of the French, and indeed
other European governments to the international
recession, and immediately launched an expansionist
economic programme. The strategy had two elements.
First it increased public expenditure to stimulate
domestic demand and, through its emphasis on redistribution to the less well off, to promote social
equality. Second, it nationalised private banks and
high technology industries to provide the resources
and a higher level of investment and to facilitate
industrial restructuring.

The fate of the French 'dash for growth' illustrated the vulnerability of an open economy trying to

expand unilaterally. The government's expansion of
public sector spending and job saving programmes
stimulated inflation. Increased domestic demand
sucked in imports; decreasing price competitiveness
and tight foreign markets prevented any offsetting
rise in exports. The result was a deepening balance
of payments deficit. At the same time the budget
deficit mounted. Following pressure on the franc in
foreign exchange markets, the government was forced
to devalue in October 1981 and March 1983, and to
introduce progressively more stringent fiscal polit-
ies and wage and price controls. By reducing demand,
by accepting the balance of payments constraint and
by aiming to reduce the budget deficit, the government acknowledged by Spring 1983 that, in the words
of the Finance Minister, *"We had to adapt ourselves
to the international environment"*.
 Nevertheless, the new policy of 'rigour' does
not mean a single-minded attack on inflation as the
cause of all evils. The government stopped short of
reducing demand to a point that would precipitate any
substantial increase in unemployment. In justifying
'rigour', it seeks to emphasise the compatibility of
fiscal discipline with a positive role for the state
in the economy, a duality unfortunately forgotten in
the headlong rush to expand and reform in 1981 and
1982. It now acknowledges the resource constraint in
public expenditure but remains committed to a belief
in welfare provision as economically efficient in
maintaining demand and to promote social equality. It
continues to insist, moreover, on the necessary role
of the expanded nationalised sector in the regeneration of industry.

Germany

 Economic policy in Germany has been far more
consistent than in most other countries. The SDP under
Schmidt responded to the recessions in the seventies
with essentially the same priorities as were formulated by the Christian Democratic governments in the
post-war reconstruction years: notably a concern to
maintain price stability. Germany was far more
successful in limiting the inflationary pressures of
the seventies - whether external (oil price increases)
or internal (wage demands). The moderation of German
trade unions in wage bargaining is often attributed
to the broad consensus in German society concerning
the evils of inflation. The same consensus is reflected in the constitutional autonomy of the central bank,
giving it an authority to pursue sound monetary

objectives independently of the federal government.
The limit on the size of the federal government's
deficit established by law is a further indicator of
the need to adhere to financial rectitude. Germans
have explained the country's post-war economic success
in terms of those values and practices. Sceptics, not
only outside Germany, have seen their survival as a
consequence of economic success, arguing that they
have not been tested by economic adversity. Certainly,
the stagnation of the German economy in the early
eighties and the growth of unemployment have polarised
economic debate concerning the policies necessary for
recovery.

The problems currently facing the German economy
stem in part at least from the very success with which
it coped with those of the mid-seventies. By responding to the first oil shock with restrictive anti-
inflation policies whilst its major trading partners
tried to sustain demand, Germany was able to increase
exports and thus maintain employment. Success in
limiting inflation gave it a competitive edge throughout the seventies. However, the widespread adoption
of the German 'model' after 1979 has dampened demand
in Germany's export markets, whilst the continued
strength of the mark and other countries' relative
success in lowering their inflation rates is losing
Germany some of its competitive advantages. Rising
unemployment in the early eighties reflects this
adverse change in Germany's environment as well as
depressed domestic demand. The problem for the mid-
eighties is that if recovery is not based on a sustained demand for exports, as has been the case in the
past, where is it to come from other than domestic
expansion? The inhibitions against this remain great.
The Schmidt government followed traditional prescriptions in adopting a restrictive fiscal policy,
which the Kohl government has maintained. Interest
rates have remained high, however, for reasons which
have less to do with domestic conditions than with
the influence of high American rates. With the mark
now the world's second reserve currency and with no
restriction on capital movements, Germany is vulnerable to capital exports. The result is that investment is sluggish.

Demands for a new approach to economic policy
began to grow in the last year of the SPD/FDP
coalition. From within the SDP the government was
subject to pressure for a more expansionist policy.
But this was a less novel development in Germany
economic debate than the prescription for recovery
presented by the government's own Minister of

Economic Affairs, the Free Democrat, Otto Lambsdorff.
Lambsdorff argued that rather than wait for an international trade revival or stimulate home demand, the
government should create a climate conducive to business investment by reducing public expenditure and by
taking measures to reduce labour market rigidities.
Such an expression of free-market sentiment was
irreconcilable with the pragmatism of the Schmidt
regime and contributed to the dissolution of the
coalition. Significantly, however, it has not formed
the basis of the Kohl government's approach to economic policy despite Lambsdorff's continued occupancy
of the Ministry of Economic Affairs. The Christian
Democratic Party has always emphasised the state's
social responsibilities; it has always valued a consensual approach to economic affairs. Its record
during its first year in office has not revealed any
significant departure from past practice. It has
continued to insist on budgetary discipline, cutting
the deficit whilst making tax concession to business
and directly aiding some sectors such as the construction industry. But whilst it has trimmed public
expenditure all round, it has not followed the Lambsdorff prescriptions of drastic cuts in social security
and in state subsidies to the public and private
sectors. The public sector has not been anathematised
by the German Christian Democrats.

Italy

It is difficult to delineate a clear-cut pattern
in Italian macro-economic policy over the past decade
because of what was described as a characteristic
incoherence in Italian economic policy-making. Nevertheless, certain broad strands do emerge. Since the
mid-seventies policy has been restrictive, with the
intention of holding down domestic demand so as to
curb inflation and to prevent the balance of payments
slipping into deep deficit. To protect Italy's relatively buoyant export industries from the tendency to
uncompetitiveness brought about by the high inflation
rate, the government has resorted to frequent devaluations. Since the early eighties the deflationary
direction to policy has been more strongly pronounced
despite rising unemployment. This last phase has been
characterised by tough monetary measures by the Bank
of Italy to which the demand-management functions of
government have largely passed. Over the past few
years, however, successive governments have adopted
a more rigorous approach to budgetary policy than was
the case throughout the seventies. This reflects

growing recognition of the urgent need to address
those domestic problems which are perceived to have
damaged Italian economic performance since the early
seventies: the chronically high inflation rate and
the soaring budget deficit.

The persistence of high wage inflation in Italy
is part of the legacy of the 'Hot Autumn' of 1969
which dramatically increased trade-union power in the
labour market. The resulting high labour costs and
low productivity had a serious effect on Italian
economic competitiveness throughout the seventies.
But it was the strengthening in 1975 of the system
linking wages to the inflation rate - the *scala
mobile* - which has effectively consolidated labour's
power. Wage indexation has protected a large part of
the labour force from the effects of inflation whilst
perpetuating an inflationary wage-price spiral. It
has, moreover, centralised the process of collective
bargaining. Reform of the *scala mobile* - aimed at
reducing the element of automatic compensation for
inflation - has been a central aim of government
policy since the early eighties. There has been some
success. Unions have become more defensive in face
of the severity of the recession, acknowledging the
need to curb inflation and conceding some modification of the *scala mobile* in January 1983. The issue
remains a major item on the political agenda, however; the Craxi government has insisted on the need
for further reforms in indexation. The union response
is to use the issue as an opportunity to bargain for
government action on unemployment and tax reform. The
situation demonstrates the extent to which Italian
government has to take account of the unions in the
formulation of general economic policy.

The other major problem confronting the Italian
government - the budget deficit - is possibly even
more intractable than wage inflation. A solution
requires radical change in the way the political
system operates. The growth of the budget deficit to
its present level of almost 16 per cent of GDP has
its roots partly in the cost of the extensive social
reforms introduced in the early seventies, partly in
the use of public expenditure to compensate for the
effect of economic recession on individuals, businesses and the hardest hit regions. The welfare
system is hard to tackle because it has been used by
the dominating party in government. The Christian
Democrats - as a means of cementing political support
from its clients. Control of expenditure has also
been made difficult by the degree of independence of
the spending institutions - local and regional govern-

ment, for example - from central government control. The possibility of matching welfare increases by raising taxes has been limited by the inefficiency of the taxation system. Thus the general problem of financing a growing public sector in a period of low growth is exacerbated in Italy by the peculiar absence of discipline in its political and administrative structure.

The deficit crisis of the early eighties has been a factor in creating a consensual recognition of the pathologies in the Italian public sector. The basis of the consensus is the absence of any marked hostility by the major political parties to the principle of the public sector playing a major role in the economy but a general awareness of the dangers inherent in its present state. The Christian Democrats have a political stake in its redistributive functions, but their abuse of this function makes it difficult for them to take the painful decisions necessary for a reform. Hence their recent willingness to concede the lead in reform to a Socialist premier. The role of Craxi in insisting on controlling the deficit and cutting back public expenditure is significant in that it points to the Socialists' perception that more effective management of the public sector is a prerequisite for its use as an equitable redistributive agency and as an engine for investment and economic expansion.

Sweden

The Swedish budget deficit is one of the largest in Europe relative to the size of the economy. Controlling it is a central feature of the Social Democratic government's economic strategy. But the government's prescription of austerity in domestic consumption, reflected in public expenditure cuts and demands for wage restraint, is balanced by an active policy for growth based on the expansion of exports. In this way the Social Democrats hope to keep unemployment at the lowest level in Europe and social welfare provision amongst the highest.

Sweden, though one of the most successful of the European economies in the post-war period, has been severely jolted since the mid-seventies. Lower growth rates have accentuated the problem of financing its extensive welfare system. Almost total dependence on imported energy, notably oil, has made its balance of payments vulnerable. Meanwhile, its export performance, on which its post-war growth was based, faltered, partly because domestic inflation left it uncom-

petitive and partly because some of its staple industries proved vulnerable to changing market conditions.
It was characteristic of the strength of the consensus on full employment that the coalitions of the non-
socialist parties in office from 1974 to 1982 felt cons-
trained to absorb rising unemployment by extensive
public works and job creation programmes. Less
characteristic was the resort to nationalisation and
subsidisation of those sectors of industry such as
steel and shipbuilding which were clearly in structural decline. A distinctive feature of the mixed
economy developed by the Swedish Social Democrats
during their forty years in office until 1976 is that
relatively little was taken into public ownership:
the recent moves were ostensibly to facilitate
rationalisation, but really to preserve jobs. This
addition to the public sector was a heavy burden to
finance. By 1980 Swedish public expenditure reached
over 60 per cent of GDP and the budget deficit was
over 10 per cent of GDP. It was at this point, with
inflation high, the balance of payments in deficit
and a major world recession imminent, that the non-
socialist government began to change its economic
stance, insisting on the need to contain public
expenditure and hold back domestic demand.
    The debate on the role of the public sector
which was provoked by the deficit crisis has been one
of the factors polarising Swedish politics. The Social
Democrats accused the non-socialists of using the
opportunity to undermine the welfare state and, in
the election campaign of 1982, sought to associate
the non-socialists' readiness to cut expenditure with
a willingness to allow unemployment to rise further.
Reference by the Conservatives, the largest part in
the non-socialist bloc, to the economic dysfunctions
of a commitment to full-employment gave some substance
to this. Nevertheless, the policies of the Social
Democrats since their return to office in 1982 demon-
strate that they, too, accept the need to come to
terms with the burden of public spending on the
economy.
    The Social Democrat's strategy of austerity and
expansionism is in certain respects in the tradition
of the party's policies of the post-war growth years.
The emphasis is on fiscal discipline by the state,
reliance on the trade unions to restrict wage levels,
a demand-management policy geared to balance of payments equilibrium, and dependence on the private
sector for the growth to sustain welfare. The circumstances of the eighties,however, make all this more
difficult. Fiscal discipline requires a combination

of higher taxation and expenditure cuts: scope for both is limited. Wage moderation may prove a wasting asset. The relationship between the Social Democratic Party and the trade union movement have historically been close, while the cohesive characters of union organisation and the traditional pattern of central bargaining with the employers has facilitated wage moderation when government asked for it. The settlement well below the inflation level in the 1983 round was an exmaple of the unions coordinating actions with government economic policy. But two factors threaten this corporate model. The first is that a price is usually demanded for cooperation: the extension of welfare benefits is no longer available; the extension of economic democracy in the form of the wage-earner funds was the basis of the 1983 bargain; but the unions are also looking to changes in taxation as part of the wage-fixing process and the government's ability to meet such non-wage demands is by no means certain for political reasons. Second, the system of bargaining by the peak organisations is changing. The employers are opting out of corporate process and insisting on industry by industry negotiations which are more sensitive to market conditions. A breakdown of centralised bargaining would result in a more fragmented, potentially competitive process in which the union movement would be less able to internalise the tensions between different sectors of the labour force so as to facilitate informal incomes policies as in the past.

The balance of payments deficits of the seventies were a product of rising energy costs which put up the import bill and industrial uncompetitiveness which hurt exports. The Social Demcorats' strategy to get back to equilibrium rests on the large devaluation made immediately after their return to office. The rise of Swedish exports during 1983 indicate that part of the strategy is working. The government has less room for manoeuvre on imports, however. Further restriction of domestic demand would increase unemployment and compromise growth targets. Currently Sweden is benefitting from the low price of oil, which constitutes a quarter of the import bill. Low oil prices reflect the depressed state of the international economy, however, and Sweden is largely dependent on international growth to expand exports. The unresolved energy problem is likely to remain a serious constraint upon the Swedish economy.

There remains also the question of Swedish industry's willingness to deliver export-led growth.

Overview

The Social Democrats have never espoused an interventionist industrial policy, trusting to the private
sector's managerial skills. Unlike French and British
socialists, they have not seen nationalisation and
publicly directed investment as the answer to industrial problems except to ease the social costs of
running down declining sectors. The response of the
engineering, automobile and chemical sectors to the
opportunities of devaluation indicates that reliance
on the capacity of Swedish industry to adapt to
changing market conditions has not been misplaced.
However, the antagonism between business and the
Social Democrats over the wage-fund proposals is
damaging business confidence and this, together with
increasingly outspoken attacks by businessmen on the
cost of the welfare state, suggest that the renewal
of domestic investment on which the government pins
its hopes may not be forthcoming.

A clear thread running through the experiences
of the five economies since the early seventies is
the effect of the international environment. They
are all open economies and trade between them increased significantly in the high growth decades. One of
the most important lessons each has learnt since the
mid-seventies is the extent of their interdependence
and the limitations on pursuing macro-economic
policies which are significantly out of line with the
others. The penalties are more severe in the case of
a maverick trying to expand in a restrictive environment. Germany experienced lower growth than it perhaps need have done because of restrictive policies
in the mid-seventies, but was not forced to conform
by the balance of payments and/or foreign exchange
problems which disciplined Britain in the mid-seventies or, more recently, France. The French unilateral
expansionist policy of 1981-2 has become an exemplary
case illustrating the consequences of unorthodoxy.
It was cited by the British Chancellor of the
Exchequer, Sir Geoffrey Howe, in May 1983 as a lesson
for British voters. He argued that the French government had found that room for manouevre did not exist
and had been forced to implement policies that the
Conservative government in Britain had been following
for years. From a very different standpoint, the
Swedish Social Democrats in defending their economic
programme in 1982 were careful to distinguish their
expansionist policy from that of the French. They

insisted that they intended introducing domestic
austerity from the outset and that their expansion
was based on export-led growth stimulated by the
aggressive early devaluation which the French Socialist government felt inhibited from undertaking.

The convergent fiscal policies in Europe reflect
governments' awareness of the constraint imposed by
this international interdependence; but it is significant that it is the Swedish and French governments
which are most anxious to expand and, therefore, most
impatient with the constraint, which are also insistent in pressing the case for coordinated international action to revive the world economy. They point to
the danger of a common pursuit by trading partners
of export-led growth and domestic austerity being
self-defeating. This advocacy of international inter-
ventionism, based on a sense of the inadequacy of
purely national responses to economic stagnation
(whether expansionist or restrictive) is a significant guide to the divergences of outlook amongst
European governments. On the one hand, the Swedish
Social Democrats and French Socialists, unwilling to
concede that full employment is an illusory goal and
active in the pursuit of growth, see fiscal rectitude
as a necessary but not sufficient step to economic
recovery. On the other hand, the British and German
governments, insistent on the priority of price
stability, emphatic on the limited economic role of
government, believe that sound recovery must be
self-generated. The Italians meanwhile stand at a
mid-way point. Restive, like the French and Swedes,
at their international dependence, but conscious,
too, of the paramount importance of getting their
own house in order.

There is a similar mixture of divergence and
convergence in approaches to the issue of the public
sector. These are more easily appreciated if a distinction is made between the role of the public sector
in the distribution of resources (welfare-state
benefits, etc.) and its role in the control of production. In relation to the former, there is general
concern about the implications of the rapid, largely
unplanned growth of public expenditure. How to
finance it, how, indeed, to control it, are perceived
as problems by all governments. There is fear that
the public sector's share of resources is limiting
the private sector's capacity to produce wealth since
it tends to be geared to consumption more than to
production: a structural feature which worries even
those who would like to see the public sector used
as an instrument to stimulate production. Few

politicians, anywhere, would be prepared to give a
precise specification of the appropriate size of the
public sector share of resources in an economy. The
main distinction is between countries whose governments evince fundamental reservations about a large
public sector and would prefer it reduced in scale,
and countries where governments, more pragmatically,
would like to control its growth in order to preserve what they conceive to be its essential functions
in ensuring a socially equitable distribution of
resources. The British Conservative government alone
would fit into the first category, though without
as yet being able to fulfil this ambition. Such a
neo-liberal approach is less strong in continental
Europe - though it finds expression amongst Swedish
Conservatives and in some sections of the German
CDU/FDP coalition. In general, however, traditions
of 'social responsibility' in the German and Italian
Christian Democrat parties inhibit its development
in those countries.

There is a much clearer pattern of divergence on
the issues of public ownership and the use of public
power to restructure industry or revive investment.
The early eighties witnessed a reaffirmation in
France of socialist belief in the political desirability and economic necessity of an extension of
public control, whilst in Britain the Conservatives
are restoring public enterprise to private ownership. Between these extremes are interesting variants of national experience. The Swedish Social
Democrats, unlike their French and British confrères,
do not see extensive nationalisation as politically
or economically necessary, but are initiating a
radical reform in the control of Swedish industry
through their wage-funds proposals. The German
government, meanwhile, has become more deeply involved in the restructuring of industry than has
previously been the case. A discernible concern
throughout the five countries, however, is that the
common practice of the late seventies of subsidising
declining industries, whether public or private,
must be discontinued because of its heavy cost and
its diversion of resources from more productive use
- an illustration of the pervasive constraint of
financial considerations on policy.

Taxation Policy

Taxation has become an increasingly salient
issue in the Europe of the eighties. Recession, by
confronting governments with the dual effect of a

constricted tax base and expanded spending commitments, has made more acute the longer-term problem of
reconciling the upward drift of public expenditure
with diminished economic growth prospects. Throughout the seventies the share of GDP taken by taxes
rose in most countries and whilst tax levels stabilised at the end of the decade, recession has led to
a resumption of the trend at an accelerated rate.
There have been no popular tax revolts in the major
European coutnries; nevertheless there is evidence
in most countries that consent to financing the
public sector is declining in the context of recession and that resentment of the tax burden is a more
pressing constraint when so widely expressed - by
unions as well as business, working class as well as
middle class. Governments of the Left as well as the
Right are having more to recognise resistance to higher
taxation amongst their voters. In the early eighties,
moreover, political constraints on governments'
taxing abilities have been supplemented by an awareness of the economic dysfunctions of some features
of current tax systems. The search for ways to ease
the tax burden has become not only a matter of
navigating politically dangerous waters but, in some
eyes, of removing obstacles to economic recovery.

Tax levels vary considerably. Of the five
countries surveyed, Sweden is the most highly taxed
with, in 1982, government revenue amounting to over
50 per cent of GDP. Then comes France at 43 per cent
of GDP, Britain at 40 per cent, Germany at 37 per
cent and Italy at 33 per cent. With the exception of
a few years at the end of the decade when Britain
slipped behind Germany, this relationship has held
since the beginning of the seventies. There have,
however, been significant differences in the rates
of growth of the various countries' tax/GDP ratios
during the period. Sweden's tax share has grown most
rapidly and evenly; Britain's the least. Britain was
the only country whose tax share remained stable in
the course of the seventies; since 1980, however, it
has risen more sharply than elsewhere. In France
and Italy, too, the rate of increase in the tax
burden has accelerated since 1980. Germany stands
out as maintaining a stable tax/GDP ratio during the
current recession. In addition to these differences
in the dynamic of the overall tax levels, there are
significant variations in the tax structures of
European countries. These, too, have changed in
different ways since the early seventies. This diversity of experience within the parameters of economic crisis and political unease means that whilst

taxation policy has become generally problematic, the specific character of the problem and the tensions it creates differ subtly between countries.

Britain

It is arguable that the salience of tax policy in Britain in the early eighties is less a result of exigent circumstances (such as the deficit problems of Sweden, Italy and France) and more a matter of the government choosing to highlight the issue. The Conservatives made the need to reduce taxes a central theme in the 1979 general election, using the economic efficiency argument that the incentive of lower taxes would encourage growth and the more political argument that people should be freer to determine their own consumption patterns. However, because of the depth of the British recession, public spending in relation to GDP has risen during the eighties and the government, by choosing to keep tight control over its deficit, has left itself limited scope to reduce taxes, whose burden as a proportion of GDP has actually risen over the past four years. Nevertheless, whilst adjusting its time scale, the government continues to stress its commitment to lowering taxes. Moreover, it has kept the tax issue on the political agenda as part of its current attacks on the size of the public sector by insisting that failure to tackle the problem of public expenditure must lead to higher levels of taxation.

The Conservative's commitment to lower taxes is electorally popular; antipathy to taxation extends into low and middle as well as high income earners. When the Labour government in the mid-seventies sought to gain support for wage moderation by reminding unions of the 'social wage' which welfare benefits represented, it won little sympathy from most unions which insisted on their members' prime concern with increasing disposable income. The paradox is that by European standards Britain is not a highly taxed country and that during the seventies the tax burden in relation to GDP was actually growing less than in most countries despite Britain's lower than average economic growth rate. British sensitivity to taxation has probably to do with the tax structure; indeed, it is likely that it is changes in the incidence of taxation which have accounted for Conservative's ability to preside over an increase in the overall tax level without incurring any electoral backlash at its failure to fulfil its commitments.

The most striking comparative feature of the

British tax structure is that, like Sweden, direct taxes have traditionally been high in relation to indirect taxes on consumption, and that direct social security contributions have been smaller than the European average. As in Sweden, too, marginal tax rates at low and upper income levels have been comparatively high. During the late sixties and early seventies, there was a marked trend for the share of direct tax to increase, a phenomenon reversed under the Labour government, which itself was beginning to recognise that a better balance between direct and indirect taxation was desirable. The Conservative's first budget cut income tax, particularly at the top income level, but compensated for this with a dramatic increase in the VAT rate. It claimed that the cut in income tax was only the first instalment in its strategy of a major reduction in the personal tax burden, then justified failure to go further during its first term of office by the exceptional economic circumstances forcing up public spending. Maintaining this commitment and keeping the income tax rate stable has probably had a lot to do with Conservative ability to retain its image as a tax-cutting party despite the rise in the overall tax burden since 1980.

The government has been able to meet the upward movement of public expenditure partly through the windfall of North Sea oil revenue, partly through allowing 'fiscal drag' to offset some of the consequences of cutting income tax, and partly through substantially increased social security contributions from employers and employees. This last development has aroused such protests, particularly from employers, that the scope for further increase in their contributions is probably limited. In general, however, the evidence of Britain in the early eighties suggests that to be seen to be, or at least to be believed to be, concerned with lowering income tax outweighs any hostile reaction to the burden being transferred elsewhere.

France

Mitterrand's acknowledgement of the Socialist government's need to take acocunt of popular resentment at the growth of the French tax burden might seem prudent in the country which witnessed one of Europe's most dramatic tax revolts - the Poujadist movement in the early fifties. More remarkable, perhaps, is the fact that since the fifties French governments have managed to contain anti-tax feeling

whilst the country remained one of the highest taxed
in Europe. Undoubtedly France's high economic growth
in the past two decades softened the impact of high
taxation. Equally relevant is the tax structure.
Direct taxes have constituted a lower proportion of
government revenue than in any other major European
country, whilst social security contributions provide one of the highest proportions. The emphasis has
thus been on the least 'visible' forms of revenue
and the resulting tax mix has been one of the least
progressive in Europe. The recent revival of popular
tax consciousness follows a significant rise in tax
levels and a jolt to the traditional tax structure
administered by the Socialist government. These
developments have in part been the result of necessity, widening deficits in the social security funds
for example, and partly due to the Socialists' desire
to shift the burden of financing their expanded public expenditure programmes towards the better-off by
raising direct taxes and making them more progressive.

During its first year in office the Socialist
government increased maximum income tax rates and
introduced a new wealth tax, as well as tightening
up measures against tax evasion and increasing social
security contributions. Willingness to break traditional inhibitions has become most evident since
the end of the expansionist phase of the government's
economic policy. The new 'rigour' involved a more
conservative fiscal policy and, as part of its new
concern with curbing budget and social security fund
deficits, the government gave a new emphasis to
direct taxes as well as increasing consumption taxes.
Measures adopted in March 1983 included reducing the
threshold of the wealth tax, a compulsory loan to the
government equal to 10 per cent of income, and a levy
of 1 per cent on taxable income.

There are clear limits to the government's freedom to manoeuvre on the tax front. Its concern with
the inflation rate requires caution in increasing
VAT and other consumption taxes. The need to foster
business confidence limits its ability to increase
employers' social security contributions and trade
union opposition to increases in employees' contributions is equally strong. Further development of
the income tax resource, however, is equally dangerous. The visibility of this tax and the very novelty
of taxes recently introduced have heightened resentment amongst middle as well as high income earners,
the former including many of the voters who put the
Socialists in power in 1981. Mitterrand was reported

to have told his cabinet in July 1983 that the
electorate in 1986 would be more impressed by a
decline in taxation than in inflation, and has publicly pledged to reduce the ratio of tax to GDP.

Germany

The taxation issue became salient in Germany in
the context of Lambsdorff's neo-liberal programme for
economic regeneration in 1982. Lamsdorff argued that
the burden on business imposed by corporate taxation
and social security contributions was one of the
obstacles thwarting new investment and called for its
reduction. But whilst there is some support for this
position within the CDU/FDP coalition, the new government has not shown itself committed to any major
reappraisal of tax policy. Compared with other
countries, in fact, the issue is less central to the
political agenda. The tax burden has not been a major
electoral issue over the past decade and there would
appear to be greater tolerance of existing taxation
levels than in some other countries. The explanation
for this may lie in the relative stability of tax
system. In the course of the sixties Germany slipped
from second to eighth place in the league of OECD
countries tax/GDP ratio as its tax levels grew more
slowly than elsewhere. During the seventies the tax
burden increased but only at an average rate. Moreover, since 1980 it has remained stable. Compared
with other countries the German tax structure is
distinctive for the relatively even share of revenue
taken by direct taxes, consumption taxes and social
security contributions. The level of income tax in
particular remained remarkably even. The most significant change in the structure which occurred during
the early seventies was the decline of consumption
taxes in the tax share relative to social security
contributions. It might be argued, therefore, that
the absence of seemingly inexorable increase in the
tax burden or conspicuous discontinuation of experience either in tax levels or the tax structure, which
might have increased popular sensitivity to the tax
issue, has contributed to the maintenance of the
German consensus on tax.

Italy

The tax problem is at the heart of the budget
deficit crisis which dominates Italian government.
Public expenditure as a proporiton of GDP is close
to the European average; yet government revenue as a

proportion of GDP ranks amongst the lowest in Europe. This might suggest considerable scope for meeting the deficit problem by increasing revenue, but there are limits to this option. The problems lie in the character of the tax structure, in social attitudes to taxation, and in the difficulties governments have encountered in modifying these. The Italian tax system underwent a major reform in the early seventies, designed to make the system more equitable and more efficient. Previously, the structure, like France, was characterised by a reliance on indirect consumption taxes and social security contributions. Evasion of both direct and indirect taxes was extensive. In 1973 a VAT system was introduced, followed in 1974 by application of PAYE systems to income tax. Since then the tax mix has changed and income tax accounts for an increased share of the tax yield. PAYE has substantially improved its effectiveness so far as wage and salary earners are concerned; however, evasion is still widespread amongst the self-employed. VAT, on the other hand, has had less of a dynamic impact upon the revenue partly because of the importance of the small-business sector in Italy where enforcement is difficult.

The other main element in the Italian structure is social security contributions levied on employers and employees. This tax weighs heavily on employers whose contribution is three times that of employees. It has declined slightly in relation to direct and indirect taxes since the mid-seventies, largely as a result of the government's concern to relieve employers of part of the burden by replacing some of their contributions through general taxation.

Government's initial response to the problem of deficit control in the late seventies was to raise taxes, both direct and indirect, and social security contributions. More recently, however, despite the worsening of the deficit, it has had to reappraise this strategy. Some of the limits to which it can be pushed are the product of the partial modernisation of the tax structure in the seventies. The increased salience of the PAYE-based income tax may have helped the revenue, but the very difficulty of avoiding it has made those most affected by it more sensitive. Trade unions have pitched wage claims higher to compensate for tax increases; they have also argued that government is exploiting the sector least able to evade tax instead of tightening up on the self-employed, where evasion is known to be widespread. These points have had to be recognised by governments which are trying to reach agreement with the unions

195

in wage moderation and revision of the *scala mobile*. Indeed, the *scala mobile* itself imposes a constraint on government's resort to increased indirect taxes since these are compensated through the indexation mechanism by wage increases; they have therefore inflationary results. There is clearly room for marginal gains in revenue from existing tax sources, but the largest unexploited field remains undeclared income and transactions. Recent Italian budgets have granted immunities to those who make substantial payments in place of previously undeclared taxes. Such limited responses hardly match the scale of the problems, however.

Sweden

As the price of their generous social welfare system Swedes carry the highest tax burden of all the OECD countries; in 1980 government tax receipts amounted to 50 per cent of GDP. This redistribution of resources from private to public consumption was a deliberate act of policy on the part of the Social Democratic governments which had held office in Sweden from the early thirties. The growth of taxation which accompanied this shift was remarkably rapid. In 1955 Sweden was exactly midway in the league of OECD country's tax burdens; by 1965 it was at the top. The social compliance which made this shift possible can be attributed partly to its coincidence with a period of rising real incomes in one of the most successful economies in Europe; but it also testifies to the success with which the Social Democrats tutored the electorate into accepting the costs of social welfare.

The steep rise in taxation, however, continued through the seventies. Government tax receipts in 1965 stood at 2 percentage points above the OECD average; in 1980 they were almost 14 points above average. By the early seventies a critical reaction was emerging. It had become clear that there were limits to the extent which 'social democratic values', however broadly and deeply permeated, could sustain consent to the burden. The time at which the tax issue became salient in public debate is significant. It antedated perceptions that developed in the mid-seventies that Sweden was entering a period of low growth. It was manifest even whilst the extension of welfare provision was in full spate. The end of growth and the break in the link between higher taxes and a better welfare system have heightened pre-existing Swedish unease over this tax system.

Overview

Although the aggregate size of the tax burden
has become a cause of general concern, more specific
discontents have focussed on the structural characteristics of the system which has several distinctive
features. Income tax is higher than in most other
countries, both in its impact on the individual and
as a proportion of the total tax take. It is also the
most steeply progressive. While employees pay less
in the way of social security contributions than in
most other European countries, the contribution of
pay-roll levies to the national tax bill are amongst
the highest and their burden falls on employers.
Corporate taxes, however, form a smaller part of
revenue than in any other OECD country; as do property taxes. A further feature of the Swedish structure is the high share of tax paid to local authorities, levied as a proportion of taxable income on
individuals and businesses.

The tax mix in Sweden has changed considerably
since the sixties. Whilst the contribution of income tax has decreased, the contribution of pay-roll
levies has increased from about 2 per cent to over
30 per cent. A similar convergence has occurred in
most other countries, but no other started from so
low a base as Sweden. An upward shift in the share
of indirect consumption taxes in the sixties was
reversed in the course of the seventies. The substantial redistribution of the tax burden from income tax to pay-roll levies was a recognition of the
political and practical need to widen the tax base,
particularly as pay-roll levies so prominent elsewhere had been neglected in Sweden. But where Sweden
continues to be distinguished from the rest of
Europe is in the extent to which employers rather
than employees shoulder this burden, though, partly
at least, it can be passed on to consumers in the
form of higher prices. Recognition of the inflationary implications is a constraint on further exploiting this source of revenue. The relatively benign
character of Swedish corporate tax policy might
appear a paradox in a country so long dominated by
Social Democrat governments. The explanation lies
partly in the Swedish Social Democrats' respect for
the private sector so long as it performs its
function in society - creating wealth - and the low
yield of corporate tax is actually a function of the
extent to which investment can be offset against
tax.

It is on the heavy burden on individuals of
income tax that most discontent has focussed. Recognition of the harmful consequences of high marginal

rates has been sufficiently widespread as to form
the basis for the consensus supporting the major
reform decided upon in 1981. The OECD calculated in
1978 that in Sweden a man with a wife and two children on average income would lose 41 per cent of a 10
per cent rise in income, compared to 19 per cent in
Germany and only 6 per cent in France. So high a
marginal rate acts as a disincentive to accepting
extra work or moving for a higher paid job, thus
hindering labour mobility. It is also held to be
responsible for contributing to wage inflation by
forcing unions to bargain for larger increases to
compensate for tax erosion. The high marginal rate
also provides a powerful incentive for Swedes to
hire-purchase a wide range of consumer goods, as well
as houses, since interest is tax-deductible. Since
this favours wealth accumulation by high income
earners in practice, it has the unintended consequence of running counter to the redistributive purpose of a progressive tax system. The tax reform
package aims at both reducing marginal rates and
limiting the tax benefits accruing to borrowers.
    The Swedish case illustrates the seriousness of
the revenue problem even in a society whose previous
consent to its exceptional tax burden was remarkable.
Restricted by the size of the budget deficit from
further borrowing, the Social Democrats have been
forced to recognise the limited scope there now
exists for increasing taxes or widening the tax base.
There are political constraints on income tax, while
pay-roll taxes pose too great a burden on business
and consumption taxes are inflationary. The desire
to prevent the tax burden increasing is also characterised by a motive which is perhaps peculiarly
Swedish: a concern that the loss of consent to the
tax burden (implicit, for example, in the 'black economy')
undermines the civic spirit which keeps society
together. Such is the Swedish belief in the value of
consensus that a tax system which threatens that
consensus is seen as needing to be reformed.

Social Policy

    Inhibited by political prudence as well as
economic conviction from narrowing their deficits by
raising taxes, governments are having to meet the
problem by cutting expenditure. The issue of controlling public expenditure is not, of course, new. In
most countries it became salient in the mid-seventies
in the context of low growth and high inflation. One
response was a tendency to reduce levels of planned

future expenditure. This usually meant reductions in
capital investment. By the early eighties, however,
the consequence of recession - falling revenues and
higher spending on subsidisation and income support
- has forced politicians to engage in or contemplate
cuts in current expenditure. This has had profound
consequences for social policy.

A decade of low economic growth, culminating in
major depression, has throughout Europe given rise to
a 'crisis of the welfare state'. Definitions of the
nature of the crisis vary. To the neo-liberal elements within the European Right, the point has been
reached where the dangerous economic consequences of
the growth of collective social welfare have finally
become manifest. Its costs, the rigidities it has
brought about in the labour market, the lack of
individual enterprise and responsibility which they
claim it has engendered, have, they argue, contributed to economic problems and thwart the ability to
overcome them. Radical cuts in welfare state provision are, in their view, necessary for economic recovery. Some Socialists, on the other hand, see the
crisis as signifying not the welfare state's strength,
but its weakness. They believe that the economic
situation is being used by opponents of the redist-
ributional effects of social welfare as an opportunity to destroy the consensus which has sustained what
they see as a chronically underfunded, if too bureau-
cratised, welfare system. Pragmatists, of Left and
Right, see the crisis more in terms of practical
exigencies. *"France"*, said the Socialist minister
Michel Rocard, is *"living at a rate of social guaran-
tees above its means"*. Such sentiments are being
expressed throughout Western Europe.

Social policy is not the only sector affected
by governments' concern to curb public expenditure
but there are particular reasons why governments are
focussing on it. Social welfare spending has over
the past decade grown more rapidly than any other
component of public expenditure. In part this has
been the result of new commitments. Sweden, Britain
and Germany introduced major pensions reforms in the
early seventies; Italy extended its health service.
Benefits were extended as new needs or especially
disadvantaged groups were identified. The most
important factor in the growth of welfare spending,
however, has been the fulfilment of existing commitments. Transfer payments such as pensions, family
allowances, sickness and employment benefits, have
been the fastest growing component in public expenditure in every country since the mid-seventies. This

is in part because of indexation of these payments and, due to the effects of economic recession, an increase in demand. Health expenditure has also risen significantly because of factors such as ageing populations increasing the demand on services.

A climate of expenditure constraint forces governments to appraise their priorities. Policies are in competition. Britain provides a clear example of the effect of a government's priorities on public expenditure patterns. Since the Conservatives came to office in 1979, spending on defence and law and order have been increased in real terms. Housing programmes have been drastically cut back, as has higher education. One area of social policy in which the Conservatives have increased spending is training programmes for young school-leavers. The Swedish Social Democrats provide another example of the re-ordering of priorities with their increase in spending on training and wage subsidies for school-leavers, and off-setting cuts in food subsidies and the overseas aid programme.

Demand-led spending such as health care, pensions and unemployment benefit pose particular problems for governments, however. Expenditure on these services can not easily be controlled, let alone cut. Because of the need to accommodate such programmes, competition between discretionary services become more fierce. Yet their very size and buoyancy makes them vulnerable, and recent developments show that they are not immune to cost-conscious governments, Left or Right.

The problem of the recent rise in social security expenditure takes a particular form in countries such as Germany, France and Italy which organise such programmes through insurance-based funds financed largely by employer and employee contributions. Unemployment insurance funds, for example, have been especially affected by the recession which increases demand for benefits whilst cutting revenue. Health insurance funds have been hit by escalating medical costs. Though contributions have been increased recently in all countries, governments have been forced to take into account the opposition to this by employers and trade unions who participate in the management of the funds. The result has been that governments have cut back some entitlements, in the case of Germany cutting unemployment benefits and in France the pensions of early retirers. The Italian government has for some time 'bailed-out' insurance funds through the state budget, and the French government proposes a similar but

permanent restructuring of the system by absorbing
the family allowance system into the national budget
where it will be financed by general taxation.
'Budgeting' social security only relocates the problem of finance however. In Britain the Conservative
government has tackled the problem by various techniques: cutting benefit levels, limiting entitlements
and de-indexation. The Swedish Social Democrats,
though restoring some of the cuts in entitlements
introduced by their predecessors, refused to take
account of the effects of devaluation in cutting
purchasing power when they re-indexed retirement
pensions in 1983. Governments everywhere are accused
by neo-liberals of doing little more than trimming at
the edges of the social security problem, while
critics to the left argue that the foundations of the
welfare state are being eroded. Clouds of rhetoric do
not make it easy to discern a markedly differentiated
pattern of response. The Conservative government in
Britain has not applied its radicalism to the social
security system or health service with the same
clarity and force it has applied to the nationalised
industries. The privatisation of hospital laundry
services has not the same significance as the privatisation of the telecommunications industry or British
Airways. At the other extreme, the Swedish Social
Democrats, whilst staunch in their defence of the
welfare state, show an awareness of the need to
count its cost. Meanwhile, the European social security systems, however problematical, continue to fulfil
their basic function of maintaining income in face of
a major economic crisis. The political sensitivity of
this, perhaps, is the reason why they are likely to
remain essentially intact.

    The principle theme which runs through this
survey is the constraint imposed on governments by
the common impact of economic recession. There is
little divergence in the main line of macro-economic
management. Taxation policies are similarly const-
rained to follow the same broad path; whilst social
welfare policies in all five countries are subject to
the predatory intervention of finance ministries
anxious to limit exchequer commitments. Such differences in the character of economic and welfare
policies as do exist relate more to future hopes than
current realities. The Socialist governments of France
and Sweden present austerity as a necessary pause in
the pursuit of greater and more equitably distributed

wealth, whilst the Conservative government of Britain is more prone to emphasise the qualitative changes to dismantle constraints on a competitive market-orientated economy and freedom of choice for individuals. Such a market-orientated perspective also prevails in Germany under the CDU/FDP coalition, though here, unlike Britain, it is tempered by a greater attachment to a consensual approach to change.

Neo-liberalism in the early eighties is the most assertive force in the intellectual climate throughout Europe concerning economic and social policy. It is more apparent in approaches to policy in Britain, however, than elsewhere. Though evident within the Right in Germany, Italy and France, it has not usurped moderating influences less inclined to dismiss the practices of the past. Only in Sweden is there a Conservative party as insistent on a radical departure from the past as is the British Conservative Party under Mrs Thatcher. In this, perhaps, it is possible to detect one of the ways in which national contexts are important variables determining policy responses. In Britain the absence of a model of post-war economic success might well be a factor explaining the willingness of politicians to explore, and the electorate to tolerate, policies presented as radical departures from collectivist, consensual policies. The German, French and Italian Right are more influenced by their ability to claim credit for previous successes in governing their countries during the high growth decades. The Swedish Right, less broadly based a party and less encumbered by the heritage of long periods in office, is freer to challenge the past.

The European Left, meanwhile, whilst forced into the defensive by the need to adhere to the disciplines of the international economic climate, has not altogether lost its innovatory capacity. In formal terms at least, the French Socialists have forced a significant shift in economic ownership from the private to the public sector. The Swedish Social Democrats, too, are bent on introducing a novel element of economic democracy into the structure of Swedish capitalism through their wage-earner funds. The distribution of power between collective and private ownership remains an issue on which policies in Europe diverge significantly.

1. Growth of Real GDP (annual percentage changes)

	1972	1973	1974	1975	1976	1977	1978	1979	1980	1981	1982
Britain	2.1	7.6	-0.9	-0.9	3.7	1.2	3.5	2.0	-2.6	-1.3	2.3
France	5.9	5.4	3.2	0.2	5.2	3.1	3.8	3.3	1.1	0.3	1.6
Germany	4.2	4.6	0.5	-1.7	5.5	3.1	3.1	4.2	1.8	-0.1	-1.0
Italy	3.2	7.0	4.1	-3.6	5.9	1.9	2.7	4.9	3.9	0.1	-0.3
Sweden	2.3	4.0	3.2	2.6	1.1	-1.6	1.8	3.8	1.7	-0.5	0.4

2. Inflation Rates (percentage change in consumer prices from previous year)

	1972	1973	1974	1975	1976	1977	1978	1979	1980	1981	1982
Britain	7.1	9.2	16.0	24.2	16.5	15.8	8.3	13.4	18.0	11.9	8.6
France	6.2	7.3	13.7	11.8	9.6	9.4	9.1	10.8	13.6	13.4	11.8
Germany	5.5	6.9	7.0	6.0	4.5	3.7	2.7	4.1	5.5	5.9	5.3
Italy	5.7	10.8	19.1	17.0	16.8	18.4	12.1	14.8	21.2	19.5	16.6
Sweden	6.0	6.7	9.9	9.8	10.3	11.4	10.0	7.2	13.7	12.1	8.6

3. Unemployment Rates (percentage of total labour force)

	1972	1973	1974	1975	1976	1977	1978	1979	1980	1981	1982
Britain	4.4	3.3	3.2	4.7	6.0	6.4	6.1	5.6	6.9	10.6	12.8
France	2.7	2.6	2.8	4.1	4.4	4.7	5.2	5.9	6.3	7.3	8.0
Germany	0.8	0.8	1.6	3.6	3.7	3.6	3.5	3.2	3.0	4.4	6.1
Italy	6.3	6.2	5.3	5.8	6.6	7.0	7.1	7.5	7.4	8.3	8.9
Sweden	2.7	2.5	2.0	1.6	1.6	1.8	2.2	2.1	2.0	2.5	3.1

4. Public Expenditure (total outlay of government as percentage of GDP; including transfer payments)

	1972	1973	1974	1975	1976	1977	1978	1979	1980	1981	1982
Britain	40.0	41.1	45.2	46.9	46.2	44.1	43.6	43.4	45.6	47.3	
France	38.3	38.5	39.7	43.5	44.0	44.2	45.2	45.5	46.4	48.9	
Germany	40.9	41.7	44.7	49.0	48.1	48.1	47.8	47.7	48.3	49.3	
Italy	38.6	37.8	37.9	43.2	42.2	42.5	46.1	45.2	46.0	50.8	
Sweden	46.4	44.9	48.1	49.0	51.9	57.9	59.6	61.1	62.0	65.3	

5. Public Expenditure (government expenditure on goods and services as percentage of GDP)

	1972	1973	1974	1975	1976	1977	1978	1979	1980	1981	1982
Britain	18.4	18.3	20.0	21.9	21.4	20.3	19.9	19.8	21.4	21.9	22.0
France	13.2	13.2	13.6	14.4	14.6	14.7	15.0	14.9	15.2	15.8	16.2
Germany	17.1	17.8	19.3	20.5	19.9	19.6	19.7	19.6	20.1	20.6	20.3
Italy	16.1	15.5	15.1	15.4	14.8	15.3	15.9	16.2	16.4	18.2	18.4
Sweden	22.7	22.7	23.2	23.8	24.9	27.5	27.9	28.3	28.8	29.3	29.3

6. Total Tax Revenue (percentage of GDP)

	1972	1973	1974	1975	1976	1977	1978	1979	1980	1981	1982
Britain	33.9	31.8	35.2	35.9	35.5	35.2	33.6	33.4	36.0	37.3	
France	35.3	35.6	36.3	37.4	39.3	39.4	39.5	41.1	42.6	42.9	
Germany	34.7	36.2	36.4	35.9	34.3	38.0	37.7	37.4	37.7	37.3	
Italy	28.5	26.2	28.3	28.9	30.2	30.8	31.2	30.2	32.8	33.7	
Sweden	42.5	41.7	42.8	43.9	48.3	50.7	51.2	49.9	49.5	51.3	

Sources: Tables 1–5 are drawn from the *OECD Economic Outlook*, December 1983, p.152, p.161, p.163, p.159, p.157 respectively. Table 6 is drawn from *Revenue Statistics of OECD Member Countries 1965–1982*, p.68.

SELECT BIBLIOGRAPHY

Britain

The two most recent attempts to integrate analysis of policy and politics are by Americans - Douglas Ashford, *Policy and Politics in Britain* (Basil Blackwell, 1981) and Samuel Beer, *Britain Against Itself* (Faber & Faber, 1982). Both are refreshing, if sometimes idiosyncratic, and pessimistic about policy-making capability. Andrew Gamble, *Britain in Decline* (Macmillan, 1981) synthesises the various explanations about decline from a left-wing perspective. J.J. Richardson and A.G. Jordan, *Governing without Consensus* (Martin Robertson, 1979) is the best statement of corporatist policy-making in what the authors call 'post-parliamentary democracy'. Henry Drucker, Patrick Dunleavy, Andrew Gamble and Gillian Peele (eds.), *Developments in British Politics* (Macmillan, 1983) includes up-to-date analyses of the economic and social policies of the Thatcher government as well as chapters on ideology, parties, government organisation and 'new issues' like feminism and the police. Richard Rose, *Understanding the United Kingdom* (Longman, 1982) opens up territorial politics and Scottish, Welsh and Northern Ireland issues. On economic policy, Michael Stewart, *Politics and Economic Policy in the United Kingdom since 1964* (Pergamon, 1978) deploys the 'Jekyll and Hyde' thesis. Joel Barnett, *Inside the Treasury* (Andre Deutsch, 1982), the memoirs of Labour's Chief Secretary to the Treasury from 1974 to 1979, is frank about political and institutional constraints. Public expenditure control is covered in Leo Pliatzky, *Getting and Spending* (Basil Blackwell, 2nd ed., 1984). The classic account of policy formation in the Whitehall culture remains Hugh Heclo and Aaron Wildavsky, *The Private Government of Public Money* (Macmillan, 2nd ed., 1981). The economic policy of the Thatcher government is outlined in Andrew Gamble's chapter in Drucker (see above); statistics are conveniently collected in 'The Economy before the Budget', *The Economist*, 26 February 1983. On social policy, Richard Berthoud and Joan Brown, *Poverty and the Development of Anti-Poverty Policy in Britain* (Heinemann, 1981) is a clear factual state-

Select Bibliography

ment of the effect (or lack of it) of policy in the main social services and on income distribution and unemployment. Frank Field (a Labour MP) has produced *Inequality in Britain: Freedom, Welfare and the State* (Fontana, 1981) which discusses the distribution of resources and welfare inside and outside the social security system; his *Policy and Politics* (Heinemann, 1982), based on his time with the Child Poverty Action Group, is a good account of the issues of the 1960s and 1970s and the relations between government and pressure groups. Nick Bosanquet's chapter in Drucker (see above) covers recent Conservative policies, and may be supplemented by Adrian Webb and Gerald Wistow, *Whither State Welfare: Policy and Implementation in the Personal Social Services 1979-80* (Royal Institute of Public Administration, 1982).

France

So far there have been very few articles or books on France in the Mitterrand Presidency though a torrent of analyses of the Great Socialist Experiment can be expected. There are new editions of three well known general books on government and politics in France, J.E.S. Hayward, *Governing France* (Weidenfeld & Nicolson, 2nd ed., 1983) is the most relevant to the themes of this book; Henry W. Ehrmann, *Politics in France* (Little, Brown & Co., 4th ed., 1983); and Vincent Wright, *The Government and Politics of France* (Hutchinson, 2nd ed., 1983). Of course, as the chapter on France in this volume makes it clear, 1981 was not the beginning of a totally new era. There is considerable continuity - of institutions, of policy, of political culture, of economic behaviour. One can therefore usefully read J.R. Frears, *France in the Giscard Presidency* (Allen & Unwin, 1981); V. Wright (ed.) *Giscard and the Giscardians* (Allen & Unwin, 1983); W. Andrews and S. Hoffmann (eds.) *The Fifth Republic at Twenty* (State University of New York Press, 1981); and the special number of *West European Politics* (Oct. 1978) devoted to France. All of these have material relevant to the issues discussed here. The chapter on France draws particularly on Stephen S. Cohen and Peter A. Gourevitch, *France in the Troubled World Economy* (Butterworths, 1982) and the massive six volume analysis by the Commission du Bilan, set up by the new government in 1981 to assess the state of the nation. It is entitled, *La France en mai 1981* (Documentation Française) and has volumes on the economy (2), social policy, education and science, and the state and citizens' rights. Finally, there is J.R. Hough, *The French Economy* (Croom Helm, 1982).

Germany

Wolfram Bickerich, *Die 13 Jahre: Bilanz der sozialliberalen Koalition* (Spiegel-Buch, 1982, also appeared as articles in *Der Spiegel* September to November 1982) reviews the achievements of the SPD/FDP governments, with chapters on economic, education and environmental policy. David Childs and Jeffrey Johnson, *West Germany: Politics and Society* (Croom Helm, 1981), includes chapters on the economy, the education system and law and order. David Conradt, "Changing German Political Culture" in G. Almond and S. Verba (eds.), *The Civic Culture Revisited* (Little, Brown & Co., 1980), discusses in detail the changes in political behaviour since the 1960s to which reference is made in this chapter. Josef Esser et al., "'Social Market' and Modernization Policy: West Germany" in K. Dyson & S. Wilks (eds.), *Industrial Crisis: A Study of the State and Industry* (Martin Robertson, 1983). Harald Gerfin, "The Unemployment Policy Discussion in Germany in the 1970s" in A. Maddison and B. Wilpstra (eds.), *Unemployment: The European Perspective* (Croom Helm, 1982). Jack Knott, *Managing the German Economy: Budgetary Politics in a Federal State* (Lexington, 1982). Manfred Konukiewitz and Hellmut Wollmann, "Physical Planning in a Federal System: the case of West Germany" in David McKay (ed.), *Planning and Politics in Western Europe* (Macmillan, 1982). Andrei Markovits (ed.), *The Political Economy of West Germany* (Praeger, 1982). Neville Johnson, *State and Government in the Federal Republic of Germany: The Executive at Work* (Pergamon Press, 2nd ed., 1983) - for the governmental system. W.E. Paterson and Gordon Smith (eds.), *The West German Model: Perspectives on a Stable State* (Frank Cass, 1981), includes chapters on economic management, foreign policy and the universities. Geoffrey Pridham, "Terrorism and the State in West Germany during the 1970s: A Threat to Stability or a Case of Political Over-Reaction?" in Juliet Lodge (ed.), *Terrorism: a Challenge to the State* (Martin Robertson, 1981). Alfred Steinherr, "German Industrial and Labour Policy and the European Community" in W. Kohl and G. Basevi (eds.), *West Germany: a European and Global Power* (Lexington, 1982).

Select Bibliography

Italy

There are few recent works available in English
on the fields covered by this chapter. As a general
introduction to the Italian political system, P.A.
Allum, *Italy: Republic Without Government?* (Weidenfeld & Nicolson, 1973) is illuminating though now
somewhat out of date. See also R. Zariski, *Italy, The
Politics of Uneven Development* (Dryden Press, 1972).
There is nothing more recent of similar quality and
breadth. On public administration, there is David
Hine's chapter in F.F. Ridley (ed.), *Government and
Administration in Western Europe* (Martin Robertson,
1979). Italian political economy is rather more fully
covered as far as macro-economic management is concerned, with K. Allen and A. Stevenson, *An Introduction to the Italian Economy* (Martin Robertson, 1974);
G. Podbielski, *Italy: Development and Crisis in the
Postwar Economy* (Clarendon Press, 1974); and D.C.
Templeman, *The Italian Economy* (Praeger, 1981). The
student who wishes to research further should find
some reward in a dredge through *Review of Economic
Conditions in Italy, Banca Nazionale del Lavoro
Quarterly Review,* and the periodic reports on Italy
by OECD.

Sweden

Olof Ruin, "Sweden in the 1970s: Policy-Making
Becomes More Difficult" in J.J. Richardson (ed.),
Policy Styles in Western Europe, provides a general
review of changes in the Swedish policy process in
the 1970s and early 1980s. A good, book-length study
of Swedish politics is T. Anton, *Administered Politics: Elite Political Culture in Sweden* (Martinus
Nijhoff, 1980). Readers who wish to monitor particular fields can do so quite easily, free of charge,
through a series of *Fact Sheets on Sweden* on virtually every aspect of public policy obtainable from
The Swedish Institute, Box 7434, Stockholm. Recent
academic studies of particular policy fields include
the following. L. Lundqvist, *The Hare and the Tort-
oise: Clean Air Policies in the United States and
Sweden* (University of Michigan Press, 1980). A.J.
Heidenheimer and N. Elvander (eds.), *The Shaping of
the Swedish Health Service* (Croom Helm, 1980).
O. Ruin, "External Control and Internal Participa-
tion: Trends in the Politics and Policies of Swedish
Higher Education" in H. Daalder and E. Shilds (eds.),
Universities, Politicians and Bureaucrats (Cambridge
University Press, 1982). Kjell Lundmark, "Management

of Industrial Crisis" in K. Dyson and S. Wilks, *Industrial Crisis* (op.cit.), provides an analysis of the way in which Sweden has tried to cope with industrial crises. A review of policy debates in Sweden is found in Bengt Ryden and Villy Bergstrom, *Sweden: Choices for Economic and Social Policy in the 1980s* (Allen & Unwin, 1982).

Overview

Two books provide full country-analyses of economic policies over the past decade: Andrew Cox (ed.), *Politics, Policy and the European Recession* (Macmillan, 1982); and Andrea Boltho (ed.), *The European Economy: Growth and Crisis* (Oxford University Press, 1982). The former is by a group of political scientists, the latter by economists. A longer perspective on post-war policies of the major European countries is provided in Peter Katzenstein (ed.), *Between Power and Plenty: Foreign Economic Policies of Advanced Industrial States* (University of Wisconsin Press, 1978). Useful single-author books which take a comparative perspective on economic issues are, N.W. Chamberlain, *Forces of Change in Western Europe* (McGraw-Hill, 1980), and C.F. Andrain, *Politics of Economic Policy in Western Democracies* (Duxbury Press, 1980). Brief factual accounts of current economic policy developments are contained in the *OECD Economic Outlook*, published in July and December each year. There are no book-length comparative studies of taxation policy. Material can be obtained from K.J. Newman, "International comparisons of taxes and social security contributions in 18 OECD countries, 1970-80" in Central Statistical Office, *Economic Trends*, December 1980; and OECD, *Long Term Trends in Tax Revenues of OECD Member Countries 1955-80* (1981). The constraints on government revenue are discussed in the OECD report *Public Expenditure Trends* (1978). Comparative social policies are less well covered than economic policies. Several European countries are covered in A.J. Heidenheimer, H. Heclo and C.T. Adams, *Comparative Public Policy: The Politics of Social Choice in Europe and America* (Macmillan, 2nd ed., 1983). *The Welfare State in Crisis* (OECD, 1981) is the report of a conference of academics and public officials on the problems confronting social policy in a period of low economic growth. On policy style generally, see Jeremy Richardson (ed.), *Policy Styles in Western Europe* (Allen & Unwin, 1982) with contributions on four of the five countries covered here.

CONTRIBUTORS

Drucker, H.M. is Senior Lecturer in Politics at the University of Edinburgh. He has published *Doctrine and Ethos in the Labour Party, Developments in British Politics (ed.)* and many articles on British politics; he edits the *Scottish Government Yearbooks*.

Frears, John is Reader in Politics at Loughborough University. He has published extensively on French government and politics, including *Political Parties and Elections in the Fifth Republic* (1977), *War Will Not Take Place* (with J.-L. Parodi, 1979) and *France in the Giscard Presidency* (1981).

Furlong, Paul is Lecturer in Politics at the University of Hull. A specialist in Italian politics, he has published on voting patterns, Catholicism and Christian Democracy, terrorism and most recently on the political economy. He is currently working on a study of the state, finance and industry in Italy.

Gustafsson, Gunnel is Associate Professor of Political Science at the University of Umeå. She is the author of *Local Government Reform in Sweden* (1980) and is currently working on a study of decentralisation in Sweden.

Parry, Richard is Lecturer in Social Administration at the University of Edinburgh, having formerly worked in the Department of the Environment. He has contributed the UK chapter to a forthcoming book on *Public Employment* and is the British member of the European University Institute's project on the future of the welfare state.

Contributors

Pridham, Geoffrey is Reader in Politics at the University of Bristol. His research fields are West European politics, especially Italy and Germany; his publications include *Christian Democracy in Western Germany* (1977), *The Nature of the Italian Party System* (1981), and *the New Mediterranean Democracies* (1984).

Richardson, Jeremy is Professor of Politics at the University of Strathclyde. He has published extensively on British and Scandinavian policy processes and is editor of *Policy Styles in Western Europe* (1982).

Ridley, F.F. is Professor of Political Theory and Institutions at the University of Liverpool and was also until recently a Visiting Professor at the College of Europe in Bruges. His publications include *Public Administration in France* (with J. Blondel), *The Study of Government* and *Government and Administration in Western Europe* (ed.); he edits *Parliamentary Affairs*.

Walters, Peter is Lecturer in Political Theory and Institutions at the University of Liverpool. He has published articles on British agricultural and planning policy. Over the last two years he has spent several months in Sweden researching economic policy.

Index

Index

Index

Index

Index

Index

Index

public protest 55;
technology, high 56-7
universities: France 67-
9; Germany 96-7; Italy
148-50

vocational training 92,
98

wages *see* pay
Wales 43-4
Weber, R. 170
welfare policy *see*
social services
Westinghouse system, 55-
6
Wilson, H. 16-18, 27-8,
30-2, 34, 36
women 21, 85

Zaire 52
Zimmermann, F. 110